Lecture Notes in Chemistry

Edited by G. Berthier M.J.S. Dewar H. Fischer
K. Fukui G.G. Hall J. Hinze H.H. Jaffé J. Jortner
W. Kutzelnigg K. Ruedenberg J. Tomasi

47

C.A. Morrison

Angular Momentum Theory Applied to Interactions in Solids

Springer-Verlag
Berlin Heidelberg GmbH

Author

C.A. Morrison
Harry Diamond Laboratories
2800 Powder Mill Road, Adelphi, MD 20783, USA

ISBN 978-3-540-18990-9   ISBN 978-3-642-93377-6 (eBook)
DOI 10.1007/ 978-3-642-93377-6

Library of Congress Cataloging-in-Publication Data. Morrison, Clyde A. (Clyde Arthur), 1926-
Angular momentum theory applied to interactions in solids / C.A. Morrison. p. cm. –(Lecture
in chemistry; 47) Includes bibliographies.

1. Crystal field theory. 2. Angular momentum. I. Title. II. Series.
QD475.M68 1988 530.4'1–dc 19 88-4454

Preface

From December 1985 through March 1986 the text of this book formed the basis of an in-hours course taught by the author at Harry Diamond Laboratories. Considerable assistance in revising and organizing the first draft was given by John Bruno.

The original draft of these notes was based on a collection of lectures delivered at the Universidade Federal de Pernambuco, Recife, Brazil, between 2 November 1981 and 2 December 1981. The visit to Recife was a response to an invitation of Professor Gilberto F. de Sá of the Physics Department. In the preparation of these notes I made many requests of my coworkers for earlier results and recollections of our early work. Among those consulted were Donald Wortman, Nick Karayianis, and Richard Leavitt. Further, a number of suggestions from my Brazilian colleagues helped make the lectures more clear. Particular among these were Professor Oscar Malta and Professor Alfredo A. da Gama both of whom I wish to thank for their help.

Encouragement and assistance with funding for much of this work came from Leon Esterowitz of the Naval Research Laboratory and Rudolph Buser and Albert Pinto of the center for Night Vision and Electro-Optics.

I would like to take this opportunity to express my thanks to my brother Robert Morrison and his wife Pat who, very early in my college life, gave me crucial financial assistance and encouragement. This assistance was given at considerable hardship to themselves. Thank you Bob and Pat! Also, for more years than I care to mention I wish to thank my helpmeet, Su, for companionship, encouragement and all those many small things that have made this possible and our life enjoyable.

CONTENTS

CONTENTS (cont'd)

CONTENTS (cont'd)

TABLES

Angular Momentum Theory Applied to Interactions in Solids

Clyde A. Morrison

Harry Diamond Laboratories, 2800 Powder Mill Road, Adelphi, MD 20783-1197

1. INTRODUCTION AND REVIEW OF PREVIOUS LITERATURE

The purpose of this report is to provide students with a means to calculate the energy levels of impurity ions in solids and analyze experimental results. In order to achieve this goal it is necessary to review material which many readers may have been exposed to elsewhere. This review is rather brief, and the bibliography (sect. 1.4) includes references to specific sections of textbooks, monographs, or papers where the material is discussed in more detail. Unfortunately, as in most fields of science, a number of different notations (in various alphabets) are used in the literature; it is hoped that this review will help overcome some of this difficulty.

Many mathematical results are presented without proof with a view towards brevity. The review of angular momentum is directed towards the use of the various tabulations of matrix elements, n-j symbols, and group tables, and our study of group theory is simply directed toward the use of the abundant tables of the 32 point groups.

Most of the discussion of crystal-field theory is devoted to the point-charge or point-multipole model. This is quite natural since most of its recent development has been done by research workers at Harry Diamond Laboratories (HDL). Very recently, research workers under the direction of Paul Caro in France have significantly extended the point-multipole crystal-field model and their contributions are also discussed.

Finally, the ongoing work at HDL on the transition-metal ions is discussed. Most of these results are unpublished at present, but computer programs are used to illustrate directly the effect of various free-ion and crystal-field interactions.

1.1 The Hydrogen Atom

We assume that we have an electron of charge $-e$, mass m, and an infinitely massive nucleus of charge Ze. The nonrelativistic Hamiltonian for this system is

$$\frac{p^2}{2m} - \frac{Ze^2}{r} \quad , \tag{1.1}$$

where p is the momentum of the electron. The time-independent Schrodinger equation for this system is given by

$$H\psi = E\psi \quad ,$$

$$\left(\frac{\hbar^2}{2m} \nabla^2 - \frac{Ze^2}{r} \right) \psi = E\psi \quad , \tag{1.2}$$

where we have used the relation $\vec{P} = -i\hbar\nabla$. The bound solutions to equation (1.2) satisfying all the boundary conditions are

$$\psi_{n\ell m} = R_{n\ell}(r)Y_{\ell m}(\theta,\phi) \quad , \tag{1.3}$$

where the $R_{n\ell}(r)$ are the associated Laguerre polynomials and the $Y_{\ell m}(\theta,\phi)$ are spherical harmonics. The radial functions $R_{n\ell}(r)$ are of little interest here and are not discussed further. The energy in equation (1.2) is

$$E_n = \frac{-z^2 me^4}{2n^2\hbar^2} \tag{1.4}$$

where $n \geq 1$. The quantum number ℓ (angular momentum) is restricted to the values

$$\ell < n , \quad \ell = 0, 1, 2 \ldots , \tag{1.5}$$

and m, frequently referred to as the magnetic quantum number, is restricted to

$$-\ell \leq m \leq \ell \quad . \tag{1.6}$$

The spectroscopic notation for a sequence of ℓ values is

$$\begin{array}{ccccccccc} \ell = & 0, & 1, & 2, & 3, & 4, & 5, & 6, & 7 \ldots \\ & s, & p, & d, & f, & g, & h, & i, & k \ldots \; ; \end{array} \tag{1.7}$$

along with the value of n the states are referred to as 1s, 2s, 2p, 3s ... with the restriction given in equation (1.5).

The spherical harmonics are given by

$$Y_{\ell m}(\theta,\phi) = N_{\ell m}P_{\ell m}(\cos\theta)e^{im\phi} \quad , \tag{1.8}$$

where

$$N_{\ell m} = (-1)^m \left[\frac{2\ell + 1}{\pi}\right]^{1/2}\left[\frac{(\ell - m)!}{(\ell + m)!}\right]^{1/2} \quad .$$

The $P_{\ell m}(\cos\theta)$ are associated Legendre polynomials and are defined as

$$P_{\ell m}(z) = \left(1 - z^2\right)^{m/2}\left(\frac{d}{dz}\right)^m P_\ell(z) \quad , \tag{1.9}$$

and the Legendre polynomials are

$$P_\ell(z) = \frac{1}{2^\ell \ell!}\left(\frac{d}{dz}\right)^\ell \left(z^2 - 1\right)^\ell \quad . \tag{1.10}$$

The definitions given in (1.9) and (1.10) are restricted to m ≥ 0. For m < 0 we have

$$Y_{\ell -m} = (-1)^m Y_{\ell m}^* \quad , \tag{1.11}$$

where $Y_{\ell m}^*$ is the complex conjugate of $Y_{\ell m}$.

The spherical harmonics are normalized so that

$$\int Y_{\ell' m'}^*(\theta,\phi) Y_{\ell m}(\theta,\phi) \ d\Omega = \delta_{\ell'\ell} \delta_{m'm} \tag{1.12}$$

where $d\Omega = \sin\theta \ d\theta \ d\phi$, and the integration covers the range $0 \le \theta \le \pi$, $0 \le \phi \le 2\pi$.

1.2 Angular Momentum Algebra

In classical mechanics, the angular momentum of a particle is defined by

$$\vec{\ell} = \vec{r} \times \vec{p} \quad . \tag{1.13}$$

Actually, we should specify that the angular momentum so defined is about a particular origin, and $\vec{r}$ is the vector distance from this origin to the particle with momentum $\vec{p}$.

If we use the commutation relations

$$\left[x_i, p_j\right] = i\hbar\delta_{ij} \tag{1.14}$$

with x_i = x, y, or z, then we can obtain the commutation rules for angular momentum,

$$\left[\ell_x, \ell_y\right] = i\hbar\ell_z \quad , \quad \left[\ell_y, \ell_z\right] = i\hbar\ell_x \quad , \quad \text{and} \quad \left[\ell_z, \ell_x\right] = i\hbar\ell_y \quad , \tag{1.15}$$

which are the basic commutation rules for the Cartesian components of the angular momentum. For convenience here we shall drop the $\hbar$ in the commutation relations. This does not mean that we drop $\hbar$ throughout; we restore $\hbar$ simply by writing the interactions involving the angular momentum so that the $\hbar$ is contained in the constants. As an example of this, consider the spin-orbit Hamiltonian

$$H_2 = \frac{1}{2m^2c^2} \frac{1}{r} \frac{\partial U}{\partial r} \ \vec{\ell}\cdot\vec{s} \quad , \tag{1.16}$$

11

with $\vec{l}$ and $\vec{s}$ having units of angular momentum $\hbar$ (the spin angular momentum, $\vec{s}$, we will discuss later). When these are written in terms of dimensionless $\vec{l}$ and $\vec{s}$, we have

$$H_2 = \frac{\hbar^2}{2m^2c^2} \frac{1}{r} \frac{\partial U}{\partial r} \vec{l} \cdot \vec{s} \quad , \tag{1.17}$$

where $\vec{l}$ and $\vec{s}$ obey the commutation rules in equation (1.15) but $\hbar = 1$.

For our purposes here, we want to use the spherical representation of $\vec{l}$, which is given by

$$l_{+1} = -\frac{1}{\sqrt{2}} (l_x + il_y) \quad ,$$

$$l_0 = l_z \quad , \tag{1.18}$$

$$l_{-1} = \frac{1}{\sqrt{2}} (l_x - il_y) \quad ,$$

and the commutation rules are

$$[l_0, l_{+1}] = l_{+1} \quad ,$$

$$[l_{-1}, l_0] = l_{-1} \quad , \tag{1.19}$$

$$[l_{+1}, l_{-1}] = -l_0 \quad .$$

The eigenfunctions of the angular momentum are the spherical harmonics, $Y_{\ell m}(\theta,\phi)$, and

$$l_0 | \ell m \rangle = m | \ell m \rangle \quad ,$$

$$\tag{1.20}$$

$$l_{\pm 1} | \ell m \rangle = \mp \frac{1}{\sqrt{2}} [(\ell \mp m)(\ell \pm m + 1)]^{1/2} | \ell, m \pm 1 \rangle \quad ,$$

where

$$| \ell m \rangle = Y_{\ell m}(\theta,\phi) \quad .$$

Frequently, we shall use the unit vector $\hat{r}$ to indicate the argument of $Y_{\ell m}$, thus:

$$Y_{\ell m}(\theta,\phi) = Y_{\ell m}(\hat{r}) \quad .$$

12

When the $Y_{\ell m}$ are wave functions such as in equation (1.20), we have

$$Y_{\ell m}(\hat{r}) = |\ell m\rangle \quad .$$

The orthogonality of the wave functions as given by equation (1.12) is

$$\langle \ell'm'|\ell m\rangle = \delta_{\ell\ell'}\delta_{mm'} \quad .$$

Further, we shall assume that the spin angular momentum, $\vec{s}$, obeys the same commutation relations as given in equation (1.19); the two-component spinor wave functions are represented by the wave function $|sm_s\rangle$, so that the single-electron wave function for orbital and spin angular momentum is

$$|\ell m_\ell\rangle|sm_s\rangle \quad . \tag{1.21}$$

The wave functions given by equation (1.21) then obey the following:

$$\ell_0|\ell m_\ell\rangle|sm_s\rangle = m_\ell|\ell m_\ell\rangle|sm_s\rangle \quad ,$$

$$(\vec{\ell})^2|\ell m_\ell\rangle|sm_s\rangle = \ell(\ell+1)|\ell m_\ell\rangle|sm\rangle \quad ,$$

$$s_0|\ell m_\ell\rangle|sm_s\rangle = m_s|\ell m_\ell\rangle|sm_s\rangle \quad , \tag{1.22}$$

$$(\vec{s})^2|\ell m_\ell\rangle|sm_s\rangle = s(s+1)|\ell m_\ell\rangle|sm_s\rangle \quad ,$$

where, of course, $s = 1/2$. A further property of the spherical harmonics is given by

$$IY_{\ell m}(\hat{r}) = (-1)^\ell Y_{\ell m}(\hat{r}) \quad , \tag{1.23}$$

where the inversion operator is $I\hat{r} = -\hat{r}$, a property that will be used frequently in our analysis. For other symmetry operations, an explicit expression for the spherical harmonics is convenient; table 1.1 is included for this purpose. While many of the interaction terms of the Hamiltonian were derived by using spherical harmonics, it is convenient to introduce the tensor operators

$$C_{\ell m}(\hat{r}) = \left(\frac{4\pi}{2\ell+1}\right)^{1/2} Y_{\ell m}(\hat{r}) \quad . \tag{1.24}$$

TABLE 1.1. SPHERICAL TENSORS, C_{nm}, IN RECTANGULAR COORDINATES

n,m	M^a	$r^n C_{nm}$	n,m	M^a	$r^n C_{nm}$
0 0	1	1			
			5 5	$-\frac{3}{16}\sqrt{7}$	$(x+iy)^5$
1 1	$-\frac{1}{\sqrt{2}}$	$x+iy$	5 4	$\frac{3}{16}\sqrt{70}$	$z(x+iy)^4$
1 0	1	z	5 3	$-\frac{1}{16}\sqrt{35}$	$(9z^2-r^2)(x+iy)^3$
			5 2	$\frac{\sqrt{210}}{5}$	$z(3z^2-r^2)(x+iy)^2$
2 2	$\sqrt{3/8}$	$(x+iy)^2$	5 1	$-\frac{1}{16}\sqrt{30}$	$(21z^4-14z^2r^2+r^4)(x+iy)$
2 1	$-\sqrt{3/2}$	$z(x+iy)$	5 0	$\frac{1}{8}$	$63z^4-70z^2r^2+15r^4$
2 0	$\frac{1}{2}$	$3z^2-r^2$			
			6 6	$\frac{\sqrt{231}}{32}$	$(x+iy)^6$
3 3	$-\frac{1}{4}\sqrt{5}$	$(x+iy)^3$	6 5	$-\frac{3\sqrt{77}}{16}$	$z(x+iy)^5$
3 2	$\sqrt{15/8}$	$z(x+iy)^2$	6 4	$\frac{3\sqrt{14}}{32}$	$(11z^2-r^2)(x+iy)^4$
3 1	$-\frac{\sqrt{3}}{4}$	$(5z^2-r^2)(x+iy)$	6 3	$-\frac{\sqrt{105}}{16}$	$z(11z^2-3r^2)(x+iy)^3$
3 0	$\frac{1}{2}$	$z(5z^2-3r^2)$	6 2	$\frac{\sqrt{105}}{32}$	$(33z^4-18z^2r^2+r^4)(x+iy)^2$
4 4	$\frac{\sqrt{70}}{16}$	$(x+iy)^4$	6 1	$-\frac{\sqrt{42}}{16}$	$z(33z^4-30z^2r^2+5r^4)(x+iy)$
4 3	$\frac{\sqrt{35}}{4}$	$z(x+iy)^3$	6 0	$\frac{1}{16}$	$231z^6-315z^4+105z^2-5$
4 2	$\frac{\sqrt{10}}{8}$	$(7z^2-r^2)(x+iy)^2$			
4 1	$-\frac{\sqrt{5}}{4}$	$z(7z^2-3r^2)(x+iy)$			
4 0	$\frac{1}{8}$	$35z^4-30z^2r^2+3r^4$			

^{a}*Multiplier to entry on the right.*

Since $Y_{\ell m}^*(\hat{r}) = (-1)^m Y_{\ell,-m}(\hat{r})$, we have

$$C_{\ell m}^*(\hat{r}) = (-1)^m C_{\ell,-m}(\hat{r}) \quad . \tag{1.25}$$

14

The use of $C_{\ell m}$ rather than $Y_{\ell m}$ in the interaction terms eliminates almost all the factors of 4π. An example of this is the coupling rule for spherical harmonics (Rose, 1957, p 61):*

$$Y_{kq}Y_{nm} = \sum_{\ell} \left[\frac{(2k+1)(2n+1)}{4\pi(2\ell+1)}\right]^{1/2} \langle k(0)n(0)|\ell(0)\rangle \; \langle k(q)n(m)|\ell(q+m)\rangle \; Y_{\ell,q+m} \quad ,$$

(1.26)

but

$$C_{kq}C_{nm} = \sum_{\ell} \langle k(0)n(0)|\ell(0)\rangle \; \langle k(q)n(m)|\ell(q+m)\rangle \; C_{\ell,q+m} \quad .$$ (1.27)

In equations (1.26) and (1.27), all the tensor operators have the same argument. The quantities in angular brackets in equations (1.26) and (1.27) are Clebsch-Gordan (C-G) coefficients, which we cover in section 2.

1.3 Problems

1. The inversion operator I converts the vector $\vec{r}$ to $-\vec{r}$. Show that

$$IY_{\ell m}(\hat{r}) = (-1)^{\ell}Y_{\ell m}(\hat{r}) \quad .$$

2. If

$$C_2(x)(x,\ y,\ z) \to (x,\ -y,\ -z) \quad ,$$

$$C_2(y)(x,\ y,\ z) \to (-x,\ y,\ -z) \quad ,$$

$$\sigma_h(x,\ y,\ z) \to (x,\ y,\ -z) \quad ,$$

$$C_2(z)(x,\ y,\ z) \to (-x,\ -y,\ z) \quad ,$$

$$C_2(1)(x,\ y,\ z) \to (y,\ x,\ -z) \quad ,$$

$$C_2(2)(x,\ y,\ z) \to (-y,\ -x,\ -z) \quad ,$$

$$\sigma(\underline{}x)(x,\ y,\ z) \to (-x,\ y,\ z) \quad ,$$

$$\sigma(\underline{}y)(x,\ y,\ z) \to (x,\ -y,\ z) \quad ,$$

_____

*References are listed, alphabetically by author, at the end of each section, along with uncited bibliographic entries pertinent to each topic.

$$\sigma(\underline{|}1)(x, y, z) \rightarrow (-y, -x, z) \quad,$$

$$\sigma(\underline{|}2)(x, y, z) \rightarrow (y, x, z) \quad,$$

show that

$$C_2(x)Y_{\ell m} = (-1)^\ell Y_{\ell -m}(\hat{r}) \quad,$$

$$C_2(y)Y_{\ell m}(\hat{r}) = (-1)^{\ell +m}Y_{\ell -m}(\hat{r}) \quad,$$

$$\sigma_h Y_{\ell m}(\hat{r}) = (-1)^{\ell +m}Y_{\ell m}(\hat{r}) \quad,$$

$$C_2(z)C_{kq} = e^{i\pi q}C_{kq} \quad,$$

$$C_2(1)C_{kq} = (-1)^k e^{i(\pi/2)q}C_{k-q} \quad,$$

$$C_2(2)C_{kq} = (-1)^k e^{-(\pi/2)q}C_{k-q} \quad,$$

$$\sigma(\underline{|}x)C_{kq} = IC_2(x)C_{kq} = C_{k-q} \quad,$$

$$\sigma(\underline{|}y)C_{kq} = IC_2(y)C_{kq} = (-1)^q C_{k-q} \quad,$$

$$\sigma(\underline{|}1)C_{kq} = e^{i(\pi/2)q}C_{k-q} \quad,$$

$$\sigma(\underline{|}2)C_{kq} = e^{-i(\pi/2)q}C_{k-q} \quad,$$

by using equations (1.8), (1.9), and (1.10), or by using table 1.1.

3. Using the generating function for $P_\ell(\cos \theta)$, obtain the result

$$\frac{1}{|\vec{r}_1 - \vec{r}_2|} = \sum \frac{r_<^\ell}{r_>^{\ell+1}} P_\ell(\cos \theta) \quad,$$

where $\cos \theta = \hat{r}_1 \cdot \hat{r}_2$ and $r_<$ is the smallest of the vectors r_1 and r_2. From the addition theory of spherical harmonics, we have

$$P_\ell(\cos \theta) = \sum_{m=-\ell}^{\ell} C_{\ell m}^*(\hat{r}_1)C_{\ell m}(\hat{r}_2) \quad.$$

The generating function for the Legendre polynomials is given in Rainville (1960) as

$$\frac{1}{\left[1 + h^2 - 2hz\right]^{1/2}} = \sum_{\ell=0}^{\infty} h^{\ell} P_{\ell}(z) \quad , \quad \text{with } |h| < 1 \text{ and } z = \cos\theta \quad .$$

4. From the generating function for $P_{\ell}(z)$, show that $P_{\ell}(1) = 1$, $P_{\ell}(-1) = (-1)^{\ell}$, and

$$\sum_{m=-\ell}^{\ell} C_{\ell m}^{*}(\hat{r}) C_{\ell m}(\hat{r}) = 1 \quad .$$

Show, in two ways, that the sum in problem 3 is $(-1)^{\ell}$ when $\vec{r}_2 = -\vec{r}_1$, for arbitrary ℓ.

5. By expanding $\hat{r} \times \vec{\ell}$ show that

$$\nabla = \hat{r}\frac{\partial}{\partial r} - \frac{i\hat{r} \times \vec{\ell}}{r}$$

which is a convenient form of the ∇ operator for spherical problems.

1.4 Annotated Bibliography and References

Condon, E. U., and H. Odabasi (1980), Atomic Structure, Cambridge University Press, Cambridge, U.K. Chapters 3 and 4 give a thorough discussion of the hydrogenic wave function. Pages 190 and 191 tabulate the radial wave functions.

Judd, B. R. (1963), Operator Techniques in Atomic Spectroscopy, McGraw-Hill, New York, NY. This outstanding textbook is frequently referred to in this report; the book is almost completely free of typographical errors. The problems in general are very difficult and require considerable time. Only the preface and the first two pages are pertinent here.

Leighton, R. (1959), Principles of Modern Physics, McGraw-Hill, New York, NY, chapter 5, The One Electron Atom.

Polo, S. R. (1961, June 1), Studies on Crystal Field Theory, Volume I--Text, Volume II--Tables, RCA Laboratories, under contract to Electronics Research Directorate, Air Force Cambridge Research Laboratories, Office of Aerospace Research, contract No. AF 19(604)-5541. [Volume II gives date as June 1, 1961.] The symmetry operations are discussed on pp 1-4ff; Clebsch-Gordon coefficients on pp 8.1ff; and excellent tables of $P_{\ell}(z)$, $P_{\ell m}(z)$, and $Y_{\ell m}$ are given in the appendix, all in Vol. I.

Rainville, E. D. (1960), Special Functions, Macmillan, New York, NY. This is just one of the numerous texts written by this outstanding teacher.

Rose, M. E. (1957), Elementary Theory of Angular Momentum, Wiley, New York, NY, chapter II.

Sobelman, I. I. (1979), Atomic Spectra and Radiative Transitions, Springer-Verlag, New York, NY. This is a very excellent monograph and we frequently refer to Sobelman's derivations. Beware of typographical errors!! Pages 1 through 12 are pertinent.

Watanabe, H. (1966), Operator Methods in Ligand Field Theory, Prentice-Hall, Englewood Cliffs, NJ. Pages 11 and 12 are applicable here; also, the introduction is interesting. We frequently refer to this monograph. Most of the equations are free of typographical errors. Tables of explicit expressions for $Y_{\ell m}$ for $0 \leq \ell \leq 6$ are given in appendix 1.2.

2. CLEBSCH-GORDAN COEFFICIENTS

For our purpose, it is convenient to define the Clebsch-Gordan (C-G) coefficients as the coefficients in the transformation from two angular momentum spaces, say, $\vec{\ell}$ and $\vec{s}$, to form the composite space $\vec{j}$. That is,

$$|jm> = \sum_{\mu} <\ell(\mu)s(m-\mu)|j(m)> \ |\ell\mu>|s,m-\mu> \ , \qquad (2.1)$$

where the quantity $<\ell(\mu)s(m-\mu)|j(m)>$ is a C-G coefficient. The limits on the sum in equation (2.1) are not given, as it is assumed (and will be assumed in the following) that the sums cover all values for which the C-G coefficient does not vanish. Since we wish to have an orthonormal basis, we have

$$<j'm'|jm> = \delta_{jj'}\delta_{mm'}$$

$$= \sum_{\mu\mu'} <\ell(\mu)s(m-\mu)|j(m)> <\ell(\mu')s(m'-\mu')|j'(m')> \qquad (2.2)$$

since

$$<\ell\mu'|\ell\mu> = \delta_{\mu\mu'}, \quad \text{and} \quad <s,m'-\mu'|s,m-\mu> = \delta_{m-\mu,m'-\mu'} \ . \qquad (2.3)$$

Thus, we have

$$\delta_{jj'} = \sum_{\mu} <\ell(\mu)s(m-\mu)|j(m)> <\ell(\mu)s(m-\mu)|j'(m)> \ , \qquad (2.4)$$

an important and very useful result. If we assume (correctly) that the same coefficients connect the $\vec{j}$ space to the $\vec{\ell}$ and $\vec{s}$ spaces, we can obtain another condition on the C-G coefficients, namely,

$$\delta_{\ell\ell'}\delta_{ss'} = \sum_{j} <\ell(m_\ell)s(m_s)|j(m_\ell+m_s)> <\ell'(m_\ell)s'(m_s)|j(m_\ell+m_s)> \ . \qquad (2.5)$$

Some other relations among C-G coefficients are

$$\langle a(\alpha)b(\beta)|c(\gamma)\rangle = 0$$

if $|\alpha| > a$, or $|\beta| > b$, or $|\gamma| > c$, $\qquad (2.6)$

and if $\gamma \neq \alpha + \beta$.

The C-G coefficients vanish unless the three angular momenta obey the triangle condition, that is, $|a - b| \leq c \leq a + b$ for any permutation of a, b, or c. Three of the most important symmetry relations of C-G coefficients are

$$\langle a(\alpha)b(\beta)|c(\gamma)\rangle = (-1)^{a+b-c} \langle a(-\alpha)b(-\beta)|c(-\gamma)\rangle$$

$$= (-1)^{a+b-c} \langle b(\beta)a(\alpha)|c(\gamma)\rangle$$

$$\qquad (2.7)$$

$$= (-1)^{a-\alpha} \left(\frac{2c + 1}{2b + 1}\right)^{1/2} \langle a(\alpha)c(-\gamma)|b(-\beta)\rangle \quad ,$$

where $\alpha + \beta = \gamma$.

The coupling procedure given in equation (2.1) can be deceptive. If we had formed the wave function $|jm\rangle$ by coupling s to ℓ as

$$|jm\rangle = \sum_{\mu} \langle s(\mu)\ell(m-\mu)|j(m)\rangle \; |s\mu\rangle|\ell,m-\mu\rangle \quad , \qquad (2.8)$$

the relation to equation (2.1) would be, from equation (2.7),

$$|jm\rangle = \sum_{\mu} (-1)^{s+\ell-j}\langle\ell(\mu)s(m-\mu)|jm\rangle \; |\ell\mu\rangle|s,m-\mu\rangle \quad ,$$

in which we have changed the summing index $\mu \to m-\mu$. This overall phase factor may have no effect, but in a long involved problem a switch from ℓ-s to s-ℓ coupling can cause errors. Algebraic expressions for $\langle a(\alpha)b(\beta)|c(\alpha+\beta)\rangle$ are given by Rose (1957, pp 224-225), for c = 1/2, 1; algebraic expressions for the related 3-j symbols for $1/2 \leq c \leq 2$ are given in Brink and Satchler (1962, p 36). A few special C-G coefficients of interest are

$$\langle a(0)b(0)|c(0)\rangle = (-1)^{s/2} \left(\frac{2c + 1}{s + 1}\right)^{1/2} \frac{T(s)}{T(s_1)T(s_2)T(s_3)} \qquad (2.9)$$

where

$$s_1 = -a + b + c, \quad s_2 = a - b + c, \quad s_3 = a + b - c, \quad s = a + b + c \quad ,$$

20

and

$$T(s) = \frac{\left(\frac{s}{2}\right)!}{\sqrt{s!}}$$

$$\langle a(\alpha)b(\beta)|a+b(\alpha+\beta)\rangle = \left[\frac{\binom{2a}{a+\alpha}\binom{2b}{b+\beta}}{\binom{2a+2b}{a+b+\alpha+\beta}}\right]^{1/2} \quad ,$$

$$\binom{a}{b} = \frac{a!}{b!(a-b)!} \quad , \tag{2.10}$$

$$\langle a(\alpha)0(0)|b(\beta)\rangle = \delta_{a,b}\delta_{\alpha,\beta} \quad ,$$

$$\langle \ell(0)k+2(0)|\ell(0)\rangle = -\frac{k+1}{k+2}\left[\frac{(2\ell+k+2)(2\ell-k)}{(2\ell-k-1)(2\ell+k+3)}\right]^{1/2} \langle \ell(0)k(0)|\ell(0)\rangle \quad .$$

The commutation relations for the spherical components of the angular momentum of a single electron given in equation (1.19) can be written compactly in terms of C-G coefficients as

$$[\ell_\mu,\ell_\nu] = \sqrt{2} \langle 1(\nu)1(\mu)|1(\mu+\nu)\rangle \ell_{\mu+\nu} \quad \text{and} \tag{2.11}$$

$$[s_\mu,s_\nu] = \sqrt{2} \langle 1(\nu)1(\mu)|1(\mu+\nu)\rangle s_{\mu+\nu} \quad . \tag{2.12}$$

The total angular orbital momentum operator for a system of N electrons is

$$\vec{L} = \sum_i^N \vec{\ell}(i) \quad , \tag{2.13}$$

and the total spin angular momentum operator is

$$\vec{S} = \sum_i^N \vec{s}(i) \quad . \tag{2.14}$$

The spherical components of these operators obey the same commutation relations as equations (2.11) and (2.12), or

$$[L_\mu,L_\nu] = \sqrt{2} \langle 1(\nu)1(\mu)|1(\mu+\nu)\rangle L_{\mu+\nu} \quad \text{and} \tag{2.15}$$

$$[S_\mu,S_\nu] = \sqrt{2} \langle 1(\nu)1(\mu)|1(\mu+\nu)\rangle S_{\mu+\nu} \quad . \tag{2.16}$$

Also, it should be noted that

$$[L_\mu, \ell_\nu(i)] = \sqrt{2} \, \langle 1(\nu)1(\mu)|1(\mu+\nu)\rangle \, \ell_{\mu+\nu}(i) \quad \text{and} \tag{2.17}$$

$$[S_\mu, s_\nu(i)] = \sqrt{2} \, \langle 1(\nu)1(\mu)|1(\mu+\nu)\rangle \, s_{\mu+\nu}(i) \quad . \tag{2.18}$$

Consequently, from using equations (2.15) through (2.18), we have

$$[L_\mu, C_{kq}(i)] = \sqrt{k(k+1)} \, \langle k(q)1(\mu)|k(q+\mu)\rangle \, C_{k,q+\mu}(i) \quad , \tag{2.19}$$

$$[J_\mu, C_{kq}(i)] = [L_\mu, C_{kq}(i)] \tag{2.20}$$

with

$$\vec{J} = \vec{L} + \vec{S} \quad .$$

The Clebsch-Gordon coefficients used here are related to the 3-j symbols by

$$\langle a(\alpha)b(\beta)|c(\gamma)\rangle = \sqrt{2c+1} \, (-1)^{-a+b-\gamma} \begin{pmatrix} a & b & c \\ \alpha & \beta & -\gamma \end{pmatrix} \tag{2.21}$$

and the symmetry conditions on the 3-j symbol can be obtained from equation (2.7). The 3-j symbols are extensively tabulated by Rotenberg et al (1969).

2.1 Problems

1. Show that

$$\vec{\ell}^2 |\ell m\rangle = \ell(\ell+1)|\ell m\rangle \quad ,$$

$$\vec{s}^2 |sm\rangle = s(s+1)|sm\rangle \quad .$$

2. Consider the interaction (spin-orbit interaction)

$$H_{s-o} = \zeta \vec{\ell} \cdot \vec{s} \quad .$$

Show that the matrix elements of this interaction using the states given in equation (2.1) are

$$\langle j'm'|H_{s-o}|jm\rangle = \frac{\zeta}{2} \, [j(j+1) - \ell(\ell+1) - s(s+1)]\delta_{j'j}\delta_{m'm} \quad .$$

Hint: $\vec{j} = \vec{\ell} + \vec{s}$; consider $\vec{j}^2$.

22

2.2 Annotated Bibliography and References

Note: Check your favorite quantum mechanics text; there may be a section on C-G coefficients or 3-j symbols.

Brink, D. M., and G. R. Satchler (1962), Angular Momentum, Clarendon Press, Oxford, U.K.

Condon, E. U., and H. Odabasi (1980), Atomic Structure, Cambridge University Press, Cambridge, U.K., appendices 2 and 3.

Condon, E. U., and G. H. Shortley (1959), The Theory of Atomic Spectra, Cambridge University Press, Cambridge, U.K. The C-G notation used by Condon and Shortley is related to that used in this report as follows: $(j_1 j_2 m_1 | j_1 j_2 j m) = \langle j_1(m_1) j_2(m_2) | j(m) \rangle$.

Edmonds, A. R. (1957), Angular Momentum in Quantum Mechanics, Princeton University Press, Princeton, NJ. Relationship of the $Y_{\ell m}$ of various authors is given on page 21. His $C_q^{(k)}$ are the same as those of Judd and are the same as our C_{kq}. The relation of the C-G coefficients to other notations is given on page 52. This is a good book--but look out! It's loaded with typographical errors.

Rose, M. E. (1957), Elementary Theory of Angular Momentum, Wiley, New York, NY, chapter III. The relation of C-G coefficients to other symbols is given on page 41. The commutation rules for $[J_\mu, T_{LM}]$ in terms of C-G are given on pages 84 and 85.

Rotenberg, M., R. Bevins, N. Metropolis, and J. K. Wooten, Jr. (1969), The 3-j and 6-j Symbols, MIT Press, Cambridge, MA.

3. WIGNER-ECKART THEOREM

The Wigner-Eckart theorem states that if we have a spherical tensor T_{kq} in the space spanned by the wave functions $|JM\rangle$, then the matrix elements are

$$\langle J'M'|T_{kq}|JM\rangle = \langle J(M)k(q)|J'(M')\rangle \langle J'\|T_k\|J\rangle \quad \text{(Rose, 1957) or}$$

$$\tag{3.1}$$

$$\langle J'M'|T_{kq}|JM\rangle = (-1)^{J'-M'}\begin{pmatrix} L' & k & L \\ -M' & q & M \end{pmatrix}(J'\|T_k\|J) \quad \text{(Judd, 1963; Wybourne, 1965).}$$

The projection (q) dependence is contained in the C-G coefficients, and the factors $\langle J'\|T_k\|J\rangle$ are called the reduced matrix elements.

If we have a mixed spherical tensor, rank κ and projection λ in spin space, and rank k and projection q in orbital space, the Wigner-Eckart theorem then is

$$\langle L'M'_L S'M'_S|T^{\kappa k}_{\lambda q}|LM_L SM_S\rangle = \langle L(M_L)k(q)|L'(M'_L)\rangle \langle S(M_S)\kappa(\lambda)|S'(M'_S)\rangle \langle L'S'\|T^{\kappa k}\|LS\rangle \ .$$

$$\tag{3.2}$$

Since the C-G coefficient is purely a geometrical factor, all the physics is contained in the reduced matrix element. The Wigner-Eckart theorem allows the extraction of the geometrical factors from many complicated matrix elements; it also serves as perhaps the main motivation for the development of Racah algebra in dealing with angular momentum states.

3.1   A Single d Electron in a Crystal Field

       As an example of the use of the Wigner-Eckart theorem, we consider a problem that is simple in tensor algebra but rather important in the spectra of impurity ions in crystals: the case of a single 3d electron in an axial crystal field. Such a system could be the doubly ionized scandium ion, Sc^{2+}, substituted for a doubly ionized constituent ion of approximately the same ionic radius. The solid could be hexagonal with nearest neighbor ions located along the c-axis and the off-axis ions too distant to have an effect on the Sc^{2+} ion.

       We assume that the remainder of the electrons on Sc^{2+} are replaced by an appropriate spherical potential. The wave functions for the system are taken as

$$\psi_{3d} = R_{3d}Y_{2m}$$

$$= R_{3d}|2m\rangle \tag{3.3}$$

and as indicated we ignore any effects of the spin in our approximation. In general, the radial function, R_{3d}, can be calculated by a numerical technique such as Hartree-Fock. The Hamiltonian for the problem we consider is

$$H_{CEF} = A_{20}r^2C_{20}(\hat{r}) + A_{40}r^4C_{40}(\hat{r}) \tag{3.4}$$

(an axial crystal field is defined as $A_{kq} = 0$, $q \neq 0$) and we shall assume that

$$B_{k0} = A_{k0}\langle r^k \rangle$$

with

$$\langle r^k \rangle = \int_0^\infty r^k R_{3d}^2(r)r^2 \, dr \quad . \tag{3.5}$$

The series in (3.4) is terminated through four-fold fields (C_{40}); odd-k terms, if present, are omitted from the problem. The matrix elements of H_{CEF} are given by

$$\langle \ell m'|H_{CEF}|\ell m\rangle = B_{20} \langle \ell m'|C_{20}|\ell m\rangle + B_{40} \langle \ell m'|C_{40}|\ell m\rangle \quad . \tag{3.6}$$

For the S_4 site in yttrium aluminum garnet (YAG), the axial point charge lattice sums are

$$A_{20} = 6355 \text{ cm}^{-1}/A^2 \quad,$$

$$A_{40} = 25,089 \text{ cm}^{-1}/A^4 \quad.$$

For Sc^{2+}

$$\langle r^2 \rangle = 1.372 \text{ A}^2,$$

$$\langle r^4 \rangle = 4.053 \text{ A}^4,$$

$$B_{20} = 8719 \text{ cm}^{-1},$$

$$B_{40} = 101,686 \text{ cm}^{-1}$$

If the site were cubic, $B_{20} = 0$ and $B_{44} = \dfrac{5}{\sqrt{70}} B_{40} = 60,769$.

By the Wigner-Eckert theorem we have, generally,

$$\langle \ell m' | C_{kq} | \ell m \rangle = \langle \ell(m)k(q) | \ell(m') \rangle \ \langle \ell \| C_k \| \ell \rangle \qquad (3.7)$$

and from (1.26) we have

$$\int Y_{\ell m'}^* Y_{kq} Y_{\ell m} \ d\Omega = \left[\frac{2k+1}{4\pi} \right]^{1/2} \langle \ell(m)k(q) | \ell(m') \rangle \ \langle \ell(0)k(0) | \ell(0) \rangle \qquad (3.8)$$

where we have rearranged the order of ℓ and k in the C-G coefficients. By using the relation of C_{kq} and Y_{kq} in equation (3.8), we have

$$\langle \ell m' | C_{kq} | \ell m \rangle = \langle \ell(m)k(q) | \ell(m') \rangle \ \langle \ell(0)k(0) | \ell(0) \rangle \qquad (3.9)$$

and from (3.7) we have the important result

$$\langle \ell \| C_k \| \ell \rangle = \langle \ell(0)k(0) | \ell(0) \rangle \quad . \qquad (3.10)$$

In general

$$\langle \ell' \| C_k \| \ell \rangle = \left[\frac{2\ell+1}{2\ell'+1} \right]^{1/2} \langle \ell(0)k(0) | \ell'(0) \rangle \quad .$$

Thus using equation (3.9) in (3.6) we have

$$\langle 2m'|H_{CEF}|2m\rangle = B_{20}\ \langle 2(m)2(0)|2(m)\rangle\ \langle 2(0)2(0)|2(0)\rangle \tag{3.11}$$

$$+ B_{40}\ \langle 2(m)4(0)|2(m)\rangle\ \langle 2(0)4(0)|2(0)\rangle\ ,$$

and we notice from the symmetry properties of the C-G coefficients (eq (2.7)) that $\langle 2(-m)k(0)|2(-m)\rangle = \langle 2(m)k(0)|2(m)\rangle$, so that the states with negative projection (m < 0) have the same matrix elements as those with positive m. Notice, also, that had we considered terms in the potential, C_{kq}, with k > 4, they would not contribute since $\langle 2(m)k(0)|2(m)\rangle = 0$ for k > 4.

The C-G coefficients in equation (3.11) can be found in Rotenberg et al (1969) (the relation of the C-G coefficients and 3-j symbols is given therein) and are

$$\langle 2(0)2(0)|2(0)\rangle = -(2/7)^{1/2}\ ,$$

$$\langle 2(0)4(0)|2(0)\rangle = (2/7)^{1/2}\ ,$$

$$\langle 2(1)2(0)|2(1)\rangle = -(1/14)^{1/2}\ ,$$

$$\langle 2(1)4(0)|2(1)\rangle = \frac{-2(2/7)^{1/2}}{3}\ ,$$

$$\langle 2(2)2(0)|2(2)\rangle = -(2/7)^{1/2}\ ,$$

$$\langle 2(2)4(0)|2(2)\rangle = \frac{-(1/14)^{1/2}}{3}\ .$$

It is perhaps easier to obtain the above C-G coefficients by using equation (2.10).

Thus the energy is given by

$$E_0 = 2B_{20}/7 + 2B_{40}/7\ , \qquad\qquad m = 0\ ,$$

$$E_{\pm 1} = B_{20}/7 - 4B_{40}/21\ , \qquad\qquad m = \pm 1\ , \tag{3.12}$$

$$E_{\pm 2} = -2B_{20}/7 + B_{40}/21\ , \qquad\qquad m = \pm 2\ .$$

27

It should be noted that the trace $E_0 + 2E_1 + 2E_2$ vanishes. This is a general feature of spherical tensors. Since the diagonal matrix elements are given by

$$\langle \ell m | C_{kq} | \ell m \rangle = \langle \ell(m)k(0) | \ell(m) \rangle \, \langle \ell \| C_k \| \ell \rangle$$

then

$$\sum_{m=-\ell}^{\ell} \langle \ell m | C_{k0} | \ell m \rangle = \langle \ell \| C_k \| \ell \rangle \sum_m \langle \ell(m)k(0) | \ell(m) \rangle \quad .$$

The C-G coefficient

$$\langle \ell(m)0(0) | \ell(m) \rangle = 1$$

$$= (-1)^{\ell-m} \sqrt{2\ell+1} \, \langle \ell(m)\ell(-m) | 0(0) \rangle$$

and

$$\langle \ell(m)k(0) | \ell(m) \rangle = (-1)^{-\ell+m} \left(\frac{2\ell+1}{2k+1} \right)^{1/2} \langle \ell(m)\ell(-m) | k(0) \rangle \quad .$$

Therefore,

$$\sum_{m=-\ell}^{\ell} \langle \ell(m)k(0) | \ell(m) \rangle = \frac{2\ell+1}{\sqrt{2k+1}} \sum_m \langle \ell(m)\ell(-m) | k(0) \rangle \, \langle \ell(m)\ell(-m) | 0(0) \rangle$$

$$= (2\ell+1)\delta_{k0} \tag{3.13}$$

$$= 0 \ (k > 0, \text{ the only values of interest here}) \quad ,$$

where we have used the orthogonality condition given in equation (2.4).

The result given in equation (3.13) is a very useful check on the calculation of the energy matrices, since it is very easy to make a mistake in the evaluation of the C-G coefficients.

In our later work we will encounter problems where the matrix elements of the Hamiltonian $\langle \ell m' | H | \ell m \rangle \neq 0$. In these cases we have a set of basis functions ϕ_m (such as $|2m\rangle$ above) and we assume

$$\psi = \sum_m a_m \phi_m \quad .$$

From Schrodinger's equation we have

$$H\psi = E\psi \quad \text{or}$$

$$\sum_m a_m H\phi_m = \sum_m a_m E\phi_m$$

and we multiply through by ϕ_m^*, and integrate to obtain

$$\sum_m a_m H_{m'm} = a_{m'} E \quad \text{or}$$

(3.14)

$$\sum_m a_m \left[H_{m'm} - E\delta_{m'm} \right] = 0 \quad ,$$

which is the secular equation for determining the energy levels, E, of a system.

All our efforts will be directed toward obtaining equation (3.14) for many electron systems. We shall use the methods of group theory and other techniques to reduce the number of components in equation (3.14) to a minimun.

In the previous example the matrix elements $H_{m'm} = 0$ (m' $\neq$ m) and the energy levels were given by H_{mm}. If we consider a tetragonal crystal field given by

$$H_{CEF} = B_{20}C_{20} + B_{40}C_{40} + B_{44}\left(C_{44} + C_{4-4}\right)$$

(3.15)

where B_{44} is real, and $B_{4-4} = B_{44}$, then the only matrix element different from those of equation (3.12) is

$$\langle 2-2|H_{CEF}|22\rangle = \frac{\sqrt{70}}{21} B_{44}$$

(3.16)

$$= H_{-22} \quad ,$$

and the secular equation for these states ($|2\pm2\rangle$) is

$$\begin{vmatrix} H_{22} - E & H_{-22} \\ H_{2-2} & H_{-2-2} - E \end{vmatrix} = 0$$

(3.17)

with $H_{-2-2} = H_{22}$ and $H_{2-2} = H_{-22}$.

Expanding the determinant in equation (3.17) gives

$$\left(E - H_{22}\right)^2 = H_{-22}^2 \quad,$$

$$E = H_{22} \pm \left|H_{-22}\right| \quad,$$

and from equations (3.12) and (3.16) we have (assuming $B_{44} > 0$)

$$E_+ = -\frac{2}{7} B_{20} - \frac{1}{21} B_{40} + \frac{\sqrt{70}}{21} B_{44} \quad,$$

$$\tag{3.18}$$

$$E_- = -\frac{2}{7} B_{20} - \frac{1}{21} B_{40} - \frac{\sqrt{70}}{21} B_{44} \quad,$$

and with the energy levels E_0 and $E_{\pm 1}$ given in equation (3.12), we have all the energy levels.

The wave functions corresponding to $E_\pm$ are

$$\psi_+ = \frac{1}{\sqrt{2}} \left[\,|22\rangle + |2\text{-}2\rangle\right]$$

$$\psi_- = \frac{1}{\sqrt{2}} \left[\,|22\rangle - |2\text{-}2\rangle\right] \quad.$$

We then have all the energy levels of a single electron in a tetragonal field. In all cases we have found five energy levels, which is the number of states of the free ion ($2\ell + 1 = 5$).

An important result can be obtained for the tetragonal crystal field if we let $B_{20} = 0$ and $B_{44} = 5B_{40}/\sqrt{70}$. This is the limit of a cubic field; in this limit, from equations (3.16) and (3.10), we have

$$E_0 = E_+ = 2B_{40}/7 \quad,$$

$$\tag{3.19}$$

$$E_{\pm 1} = E_- = -4B_{40}/21 \quad.$$

The doubly degenerate level (E_0, E_+) is denoted E and the triply degenerate level $(E_{\pm 1}, E_-)$ is denoted T_2. These labels are for the irreducible representation of the cubic group in the Mulligan notation; E is referred to as Γ_3 and

T_2 as Γ_5 in the Bethe notation. The difference between the two sets of energy levels given in equation (3.17) is frequently referred to as 10Dq and is

$$10Dq = 10B_{40}/21 \quad or$$

$$B_{40} = 21Dq \quad .$$

(3.20)

The quantity Dq or 10Dq is frequently reported as an experimentally determined parameter in papers on optical data taken on transition-metal ions containing d electrons whether in cubic sites or not. The relation given in equation (3.20) holds for the many-electron configuration nd^N. The various coefficients such as Dq which are commonly used for other symmetries are given by Konig and Kremer (1977).

3.2 Problems

1. A spherical component of the angular momentum of an electron, ℓ_α, has the matrix elements

$$\langle \ell m' | \ell_\alpha | \ell m \rangle = \langle \ell(m)1(\alpha) | \ell(m') \rangle \langle \ell \| \ell \| \ell \rangle$$

from the Wigner-Eckart theorem. Also we have $\ell_0 | \ell m \rangle = m | \ell m \rangle$. Using a table of C-G coefficients (see Rose, 1957, appendix), evaluate $\langle \ell(m)1(0) | \ell(m) \rangle$ and then show that

$$\langle \ell \| \ell \| \ell \rangle = \sqrt{\ell(\ell+1)} \quad .$$

2. The tensor T_{kq} has the property $T_{kq}^* = (-1)^q T_{k-q}$. Consider the matrix element

$$\langle \ell'm' | T_{kq} | \ell m \rangle$$

and its hermitian conjugate

$$\left(\langle \ell'm' | T_{kq} | \ell m \rangle \right)^\dagger = \langle \ell m | T_{kq}^* | \ell'm' \rangle \quad .$$

By using the Wigner-Eckart theorem on the above matrix element and its hermitian conjugate show that

$$\langle \ell \| T_k \| \ell' \rangle = (-1)^{\ell-\ell'} \left[\frac{2\ell' + 1}{2\ell + 1} \right]^{1/2} \langle \ell' \| T_k \| \ell \rangle \quad .$$

31

In many books the reduced matrix elements are written $(\ell'|T_k|\ell)$ with

$$(\ell|T_k|\ell') = \sqrt{2\ell+1} \, \langle\ell|T_k|\ell'\rangle$$

so that

$$(\ell|T_k|\ell') = (-1)^{\ell-\ell'}(\ell'|T_k|\ell) \quad .$$

3. An electron is trapped at a negative ion vacancy site in a solid. Taking the effective potential the electron sees as

$$H = B_{20}C_{20} + B_{22}\left(C_{22} + C_{2-2}\right) \quad,$$

calculate the energy levels of the p state $(Y_{1m}(\hat{r}))$ of the electron. (The C-G coefficients can be found on page 225 of Rose, 1957). You will need

$$\langle\ell m|C_{20}|\ell m\rangle = \langle\ell(m)2(0)|\ell(m)\rangle \, \langle\ell(0)2(0)|\ell(0)\rangle \quad,$$

$$\langle 11|C_{22}|1-1\rangle = \langle 1(-0)2(2)|1(1)\rangle \, \langle 1(0)2(0)|1(0)\rangle \quad,$$

$$\langle 1(0)2(0)|1(0)\rangle = \frac{\sqrt{2}}{5} \quad,$$

$$\langle 1(\pm1)2(0)|1(\pm1)\rangle = \frac{1}{\sqrt{10}} \quad,$$

$$\langle 1(1)2(2)|1(1)\rangle = \frac{\sqrt{3}}{5} \quad .$$

The answer can be obtained from

$$\langle 10|H|10\rangle = \frac{2}{5} B_{20} \quad,$$

$$\langle 1\pm1|H|1\pm1\rangle = -\frac{1}{5} B_{20} \quad,$$

$$\langle 11|H|1-1\rangle = -\frac{\sqrt{6}}{5} B_{22} \quad .$$

4. By using the results given on page 101 of Ballhausen (1962) in equations (3.12) and (3.16), show that

$$B_{20} = -7Ds \quad ,$$

$$B_{40} = 21(Dq - Dt) \quad ,$$

$$B_{44} = \frac{3}{2}\sqrt{70}Dq \quad .$$

Show that for a crystal-field interaction of C_3 symmetry

$$B_{20} = -7D\sigma \quad ,$$

$$B_{40} = -14Dq - 21D\tau \quad ,$$

$$B_{43} = 2\sqrt{70}Dq \quad ,$$

as given on page 104 of Ballhausen (note that the term involving B_{44} in eq (3.15) is replaced by $B_{43}(C_{43} - C_{4-3})$).

3.3 Annotated Bibliography and References

Ballhausen, C. J. (1962), Introduction to Ligand Field Theory, McGraw-Hill, New York, NY.

Brink, D. M., and G. R. Satchler (1962), Angular Momentum, Clarendon Press, Oxford, U.K.

Hufner, S. (1978), Optical Spectra of Transparent Rare Earth Compounds, Academic Press, New York, NY.

Judd, B. R. (1963), Operator Techniques in Atomic Spectroscopy, McGraw-Hill, New York, NY, chapters 1, 2, 3, and 4.

Konig, E., and S. Kremer (1977), Ligand Field Energy Diagrams, Plenum Press, New York, NY. Pages 19 through 22 give relationships of the various notations used to describe the crystal field.

Merzbacher, E. (1961), Quantum Mechanics, Wiley, New York, NY.

Rose, M. E. (1957), Elementary Theory of Angular Momentum, Wiley, New York, NY.

Rotenberg, M., R. Bevins, N. Metropolis, and J. K. Wooten, Jr. (1969), The 3-j and 6-j Symbols, MIT Press, Cambridge, MA.

Schiff, L. I. (1968), Quantum Mechanics, 3rd ed., McGraw-Hill, New York, NY.

Sobelman, I. I. (1979), Atomic Spectra and Radiative Transitions, Springer-Verlag, New York, NY.

Watanabe, H. (1966), Operator Methods in Ligand Field Theory, Prentice-Hall, Englewood Cliffs, NJ.

Wybourne, B. G. (1965), Spectroscopic Properties of Rare Earths, Wiley, New York, NY.

4. UNIT SPHERICAL TENSORS

4.1 Discussion

Because of the power of the Wigner-Ekhart theorem, it occurred to Racah to cast the various operators representing the interactions in terms of universal quantities that could be tabulated for a frequently used many-particle system. Toward this end, Racah introduced the unit spherical tensors for the electronic configuration $n\ell^N$, which we define as

$$\langle \ell'm' | u_{kq} | \ell m \rangle = \langle \ell(\mu)k(q) | \ell(m') \rangle \, \delta_{\ell\ell'}$$

for the orbital space and

$$\langle \ell'm's'm'_s | v^{\kappa k}_{\lambda q} | \ell m s m_s \rangle = \langle \ell(m)k(q) | \ell(m') \rangle \, \langle s(m_s)\kappa(\lambda) | s(m'_s) \rangle \, \delta_{\ell\ell'} \delta_{ss'} \quad ,$$

$$(4.1)$$

for orbital and spin space.

The generalization to an N-electron system is simply

$$U_{kq} = \sum_{i}^{N} u_{kq}(i) \quad \text{and}$$

$$(4.2)$$

$$V^{\kappa k}_{\lambda q} = \sum_{i}^{N} v^{\kappa k}_{\lambda q}(i) \quad .$$

A simple and often used example of these tensors in orbital space is

$$\sum_i C_{kq}(i) = \sum_i \langle \ell | C_k | \ell \rangle \, u_{kq}(i)$$

$$= \langle \ell | C_k | \ell \rangle \, U_{kq} \quad , \tag{4.3}$$

where

$$\langle \ell | C_k | \ell \rangle = \langle \ell(0) k(0) | \ell(0) \rangle$$

(we omit the upper limit on the i sum in the remainder of the discussion). The angular momentum is simply related to unit tensors by

$$L_\mu = \sum_i \ell_\mu(i) = \sum_i \langle \ell | \ell | \ell \rangle \, u_{1\mu}(i)$$

$$L_\mu = \langle \ell | \ell | \ell \rangle \, U_{1\mu} \quad , \quad \text{and} \tag{4.4}$$

$$\langle \ell | \ell | \ell \rangle = \sqrt{\ell(\ell+1)} \quad .$$

An example of a tensor in a mixed spin and orbital space occurs in the hyperfine interaction H_5, given by

$$H_5 = (2\beta\beta_N\mu_N/I) \sum_i \vec{N}_i \cdot \vec{I}/r_i^3 \quad , \tag{4.5}$$

where β is the Bohr magneton, β_N is the nuclear magneton, μ_N is the nuclear moment, and I is the nuclear spin. Now

$$\vec{N}_i = \vec{\ell}_i - \vec{s}_i + 3\vec{r}_i(\vec{r}_i \cdot \vec{s}_i)/r_i^2 \tag{4.6}$$

or

$$N_q(i) = \ell_q(i) - \sqrt{10} \sum_\nu \langle 1(\nu)2(q-\nu) | 1(q) \rangle \, s_\nu(i) C_{2,q-\nu}(i) \tag{4.7}$$

(we show in sect. 6 how eq (4.7) is obtained from eq (4.6)).

The part of $N_q(i)$ containing $\ell_q(i)$ can be written in terms of U_{1q}, as in the second part of equation (4.4):

$$\sum_i s_\nu(i) C_{2,q-\nu}(i) = \langle s \| s \| s \rangle \langle \ell \| C_2 \| \ell \rangle \, V^{1\,2}_{\nu,q-\nu} \quad . \tag{4.8}$$

A component of $\vec{N} = \sum_i \vec{N}_i$ can be written

$$N_q = \sqrt{\ell(\ell+1)} \, U_{1q} - \sqrt{10} \, \sqrt{s(s+1)} \, \langle \ell(0)2(0) | \ell(0) \rangle$$

$$\tag{4.9}$$

$$\times \sum_\nu \langle 1(\nu)2(q-\nu) | 1(q) \rangle \, V^{1\,2}_{\nu,q-\nu} \quad .$$

Thus, equation (4.5) can be written

$$H_5 = \left(2\beta\beta_N \mu_N / I \right) \langle 1/r^3 \rangle \sum_q N_q I^*_q \quad , \tag{4.10}$$

with N_q given by equation (4.9).

Various authors use different normalizations of the unit spherical tensors. The relation of the spherical tensors used here to those tabulated by Nielson and Koster (1963) and by Polo (1961) are

$$\langle L'S'\alpha' \| U_k \| LS\alpha \rangle = (L'S\alpha' \| U^{(k)} \| LS\alpha) \, \frac{\sqrt{2\ell+1}}{\sqrt{2L'+1}} \tag{4.11}$$

$$\langle L'S'\alpha' \| V^{kk} \| LS\alpha \rangle = (L'S'\alpha' \| V^{kk} \| LS\alpha) \, \frac{\sqrt{4(2\ell+1)}}{[3(2L'+1)(2S'+1)]^{1/2}} \quad . \tag{4.12}$$

In addition Polo tabulates

$$(L'S\alpha' \| C^{(k)} \| LS\alpha)$$

where

$$C^{(K)}_q = \sum_i C_{Kq}(i) \quad . \tag{4.13}$$

Nielson and Koster (1963) tabulate the reduced matrix elements of V^{11} only; the reduced matrix elements for the electronic nd^N configuration of V^{12}, V^{13}, and V^{14} are calculated by Wai-Kee Li (1971).

37

4.2 Bibliography and References

Li, Wai-Kee (1971), Reduced Matrix Elements of $V^{(12)}$, $V^{(13)}$, and $V^{(14)}$ for d^n Configurations, Atomic Data 2, 263.

Nielson, C. W., and G. F. Koster (1963), Spectroscopic Coefficients for the p^n, d^n, and f^n Configurations, MIT Press, Cambridge, MA.

Polo, S. R. (1961, June 1), Studies on Crystal Field Theory, Volume I--Text, Volume II--Tables, RCA Laboratories, under contract to Electronics Research Directorate, Air Force Cambridge Research Laboratories, Office of Aerospace Research, contract No. AF 19(604)-5541. [Volume II gives the date as June 1, 1961.]

Slater, J. C. (1960), Quantum Theory of Atomic Structure, Volume II, McGraw-Hill, New York, NY, appendix 26 and chapter 22.

5. RACAH COEFFICIENTS

The Racah coefficients arise in the coupling of three angular momenta (Rose, 1957, p 107) to form a final resultant. In the coupling of the angular momenta, we consider two coupling schemes:

$$\text{scheme A:} \quad \vec{J}_1 + \vec{J}_2 = \vec{J}_{12} \quad , \quad \vec{J}_{12} + \vec{J}_3 = \vec{J} \quad , \tag{5.1}$$

$$\text{scheme B:} \quad \vec{J}_1 + \vec{J}_3 = \vec{J}_{13} \quad , \quad \vec{J}_{13} + \vec{J}_2 = \vec{J} \quad . \tag{5.2}$$

Coupling scheme A is represented by the wave function

$$|A\rangle = \sum_{m_1 m_2 m_3} \langle j_1(m_1)j_2(m_2)|j_{12}(m_1+m_2)\rangle \langle j_{12}(m_1+m_2)j_3(m_3)|j(m)\rangle$$

$$\times |j_1 m_1 j_2 m_2 j_3 m_3\rangle \quad ; \tag{5.3}$$

scheme B is represented by the wave function

$$|B\rangle = \sum_{m_1 m_2 m_3} \langle j_1(m_1)j_3(m_3)|j_{13}(m_1+m_3)\rangle \langle j_{13}(m_1+m_3)j_2(m_2)|j(m)\rangle$$

$$\times |j_1 m_1 j_2 m_2 j_3 m_3\rangle \quad . \tag{5.4}$$

The coupling schemes A and B are connected by a unitary transformation

$$|B\rangle = \sum_{A} \langle A|B\rangle |A\rangle \quad ; \tag{5.5}$$

the coefficients of the unitary transformation are determined by taking the inner product of equation (5.3) with equation (5.4).

We define the Racah coefficients as follows:

$$W\left(j_1 j_{12} j_{13} j_3 ; j_1 j\right) = \frac{1}{\left[\left(2j_{12}+1\right)\left(2j_{13}+1\right)\right]^{1/2}} \langle A|B\rangle \quad . \tag{5.6}$$

Thus,

$$\left[\left(2j_{12}+1\right)\left(2j_{13}+1\right)\right]^{1/2} W\left(j_2 j_{12} j_{13} j_3 ; j_1 j\right)$$

$$= \sum_{m_1 m_2} \langle j_1(m_1) j_2(m_2)|j_{12}(m_1+m_2)\rangle \ \langle j_{12}(m_1+m_2) j_3(m-m_1-m_2)|j(m)\rangle$$

$$\tag{5.7}$$

$$\times \ \langle j_1(m_1) j_3(m-m_1-m_2)|j_{13}(m-m_2)\rangle \ \langle j_{13}(m-m_2) j_2(m_2)|j(m)\rangle \quad .$$

The following equation can be obtained from equation (5.7):

$$\langle j_2(m_2) j_1(m_1)|j_{12}(m_1+m_2)\rangle \ \langle j_{12}(m_1+m_2) j_3(m-m_1-m_2)|j(m)\rangle$$

$$= \sum_{j_{13}} \left[\left(2j_{12}+1\right)\left(2j_{13}+1\right)\right]^{1/2} W\left(j_2 j_1 j j_3 ; j_{12} j_{13}\right) \tag{5.8}$$

$$\times \ \langle j_1(m_1) j_3(m-m_1-m_2)|j_{13}(m-m_2)\rangle \ \langle j_2(m_2) j_{13}(m-m_2)|j(m)\rangle \quad ,$$

which is a relationship used often in our analysis. For clarity we rewrite
(5.8) as

$$\langle a(\alpha)b(\beta)|e(\alpha+\beta)\rangle \; \langle e(\alpha+\beta)d(\delta-\alpha-\beta)|c(\delta)\rangle$$

(5.9)

$$= \sum_f \sqrt{(2f+1)(2e+1)} \; W(abcd;ef) \; \langle b(\beta)d(\delta-\alpha-\beta)|f(\delta-\alpha)\rangle \; \langle a(\alpha)f(\delta-\alpha)|c(\delta)\rangle$$

(Rose, 1957). The Racah coefficient is related to the symmetrized "6-j"
symbol by the following equation:

$$W(abcd;ef) = (-)^{a+b+c+d} \begin{Bmatrix} a & b & e \\ d & c & f \end{Bmatrix} \; .$$

(5.10)

The symmetry of the "6-j" symbol is

$$\begin{Bmatrix} j_1 & j_2 & j_3 \\ \ell_1 & \ell_2 & \ell_3 \end{Bmatrix} = \begin{Bmatrix} j_2 & j_1 & j_3 \\ \ell_2 & \ell_1 & \ell_3 \end{Bmatrix} = \begin{Bmatrix} j_1 & j_3 & j_2 \\ \ell_1 & \ell_3 & \ell_2 \end{Bmatrix} = \begin{Bmatrix} j_1 & \ell_2 & \ell_3 \\ \ell_1 & j_2 & j_3 \end{Bmatrix}$$

(5.11)

and all combinations of the relations in equation (5.7). The four triads
$(j_1 \; j_2 \; j_3)$, $(j_1 \; \ell_2 \; \ell_3)$, $(\ell_1 \; j_2 \; \ell_3)$, and $(\ell_1 \; \ell_2 \; j_3)$ must be able to form a
triangle. That is,

$$|j_1 - j_2| \leq j_3 \leq j_1 + j_2 \; ,$$

(5.12)

with similar relations for the other triads.

An example of the use of Racah coefficients is in the calculation of
single-electron matrix elements of the operator

$$E_{k'q} = \sum_\lambda \langle k(q-\lambda)1(\lambda)|k'(q)\rangle \; C_{k,q-\lambda} \ell_\lambda \; ,$$

(5.13)

which arises in numerous applications. We consider the matrix element

$$\langle \ell'm'|E_{k'q}|\ell m\rangle = \langle \ell(m)k'(q)|\ell'(m')\rangle \; \langle \ell'\|E_{k'}\|\ell\rangle$$

(5.14)

by applying the Wigner-Eckart theorem, equation (3.1). Also, by taking the same matrix element of equation (5.12) we have

$$\langle \ell'm'|E_{k'q}|\ell m\rangle = \sum_\lambda \langle k(q-\lambda)1(\lambda)|k'(q)\rangle \langle \ell'm'|C_{k,q-\lambda}\ell_\lambda|\ell m\rangle \quad . \qquad (5.15)$$

Now we further consider the matrix element in equation (5.14) to obtain

$$\langle \ell'm'|C_{k,q-\lambda}\ell_\lambda|\ell m\rangle = \sum_{\ell''m''} \langle \ell'm'|C_{k,q-\lambda}|\ell''m''\rangle \langle \ell''m''|\ell_\lambda|\ell m\rangle \quad , \qquad (5.16)$$

where we have used matrix algebra on the product of two operators. If we apply the Wigner-Eckart theorem to the last matrix element in equation (5.16), we obtain

$$\langle \ell''m''|\ell_\lambda|\ell m\rangle = \langle \ell(m)1(\lambda)|\ell'(m'')\rangle \delta_{\ell\ell''} \langle \ell\|\ell\|\ell\rangle \quad ;$$

also, $m'' = m + \lambda$ as required by the C-G coefficient. We have previously shown that

$$\langle \ell''\|\ell\|\ell\rangle = \sqrt{\ell(\ell+1)} \, \delta_{\ell''\ell} \quad . \qquad (5.17)$$

Therefore,

$$\langle \ell''m''|\ell_\lambda|\ell m\rangle = \langle \ell(m)1(\lambda)|\ell''(m+\lambda)\rangle \delta_{\ell\ell''}\sqrt{\ell(\ell+1)} \quad . \qquad (5.18)$$

Using these results in the remaining C-G coefficient in equation (5.16), we have

$$\langle \ell'm'|C_{k,q-\lambda}|\ell(m+\lambda)\rangle = \langle \ell(m+\lambda)k(q-\lambda)|\ell'(m')\rangle \langle \ell'\|C_k\|\ell\rangle \quad . \qquad (5.19)$$

Substituting the result of equations (5.19) and (5.18) into equation (5.16), we have

$$\langle \ell'm'|C_{k,q-\lambda}\ell_\lambda|\ell m\rangle = \sqrt{\ell(\ell+1)}\langle \ell'\|C_k\|\ell\rangle \langle \ell(m)1(\lambda)|\ell(m+\lambda)\rangle$$
$$\times \langle \ell(m+\lambda)k(q-\lambda)|\ell'(m')\rangle \quad , \qquad (5.20)$$

giving the matrix element in equation (5.15). If we substitute the result of equation (5.20) into equation (5.15), then we have

$$\langle \ell'm'|E_{k'q}|\ell m\rangle = \sqrt{\ell(\ell+1)} \, \langle \ell'\|C_k\|\ell\rangle S \quad , \qquad (5.21)$$

where

$$S = \sum_\lambda \langle k(q-\lambda)1(\lambda)|k'(q)\rangle \langle \ell(m)1(\lambda)|\ell(m+\lambda)\rangle \langle \ell(m+\lambda)k(q-\lambda)|\ell'(m')\rangle \quad .$$

(5.22)

The last two C-G coefficients in equation (5.22) can be recoupled by using equation (5.8) or

$$\langle \ell(m)1(\lambda)|\ell(m+\lambda)\rangle \langle \ell(m+\lambda)k(q-\lambda)|\ell'(m')\rangle$$

(5.23)

$$= \sum_f \sqrt{(2f+1)(2\ell+1)}\; W(\ell 1\ell'k;\ell f)\; \langle \ell(\lambda)k(q-\lambda)|f(q)\rangle \langle \ell(m)f(q)|\ell'(m')\rangle \quad .$$

The C-G coefficients in equation (5.23) can be rearranged by using the symmetry rules of equation (2.7) to give

$$\langle 1(\lambda)k(q-\lambda)|f(q)\rangle = (-1)^{1+k-f}\; \langle k(q-\lambda)1(\lambda)|f(q)\rangle \quad . \tag{5.24}$$

This C-G coefficient and the first C-G coefficient in equation (5.22) are the only two C-G coefficients containing λ, so that

$$\sum_\lambda \langle k(q-\lambda)1(\lambda)|k'(q)\rangle \langle k(q-\lambda)1(\lambda)|f(q)\rangle = \delta_{fk'} \tag{5.25}$$

because of the orthogonality, as shown in equation (2.4), of the C-G coefficients. Thus, we get

$$S = (-1)^{1+k-k'}\sqrt{(2k'+1)(2\ell+1)}\; W(\ell 1\ell'k;\ell k')\; \langle \ell(m)k'(q)|\ell'(m')\rangle \quad , \tag{5.26}$$

which when substituted into equation (5.21) gives

$$\langle \ell'm'|E_{k'q}|\ell m\rangle = (-1)^{1+k-k'}\sqrt{\ell(\ell+1)(2\ell+1)(2k'+1)}\; W(\ell 1\ell'k;\ell k')\; \langle \ell'\|C_k\|\ell\rangle$$

(5.27)

$$\times\; \langle \ell(m)k'(q)|\ell'(m')\rangle \quad .$$

Upon comparing the result given in equation (5.14) with equation (5.27), we have

$$\langle \ell'\|E_k\|\ell\rangle = (-1)^{1+k-k'}[\ell(\ell+1)(2\ell+1)(2k'+1)]^{1/2}\; \langle \ell'\|C_k\|\ell\rangle\; W(\ell 1\ell'k;\ell k') \quad ,$$

(5.28)

43

which is a useful relation if we wished to express the tensor $E_{k'q}$ in terms of unit spherical tensors; in that case we would specialize equation (5.28) to $\ell' = \ell$ and simply replace C_{kq} in equation (4.3) by E_{kq} with the reduced matrix element given by equation (5.28). We shall have frequent occasion to express our results in terms of Racah coefficients by using equation (5.8).

5.1 Problems

It is frequently convenient to build spherical tensors of higher rank by coupling products of angular momentum operators. One such tensor is

$$T_{kq} = \sum_{\alpha} \langle 1(\alpha)1(q-\alpha)|k(q)\rangle \, \ell_{\alpha}\ell_{q-\alpha} \quad , \tag{a}$$

where

$$T^{+}_{kq} = (-1)^{q} T_{k-q} \quad ,$$

$$\left(\ell_{\alpha}\ell_{q-\alpha}\right)^{+} = (-1)^{q-\alpha}\ell_{-q+\alpha}(-1)^{\alpha}\ell_{-\alpha} \quad ,$$

in which, by the properties of the C-G coefficient, k is restricted to $0 \leq k \leq 2$. The T_{kq} thus constructed is patently a spherical tensor of rank k, projection q. The application of the Wigner-Eckart theorem (eq (3.1)) gives

$$\langle \ell m'|T_{kq}|\ell m\rangle = \langle \ell(m)k(q)|\ell(m')\rangle \, \langle \ell\|T_{k}\|\ell\rangle \tag{b}$$

with $m' = m + q$.

Direct calculation of the matrix elements gives

$$\langle \ell m'|T_{kq}|\ell m\rangle = \sum_{\alpha} \langle 1(\alpha)1(q-\alpha)|k(q)\rangle \, \langle \ell m'|\ell_{\alpha}\ell_{q-\alpha}|\ell m\rangle \quad . \tag{c}$$

By using the rules for matrix multiplication we have

$$\langle \ell m'|\ell_{\alpha}\ell_{q-\alpha}|\ell m\rangle = \sum_{\ell''m''} \langle \ell m'|\ell_{\alpha}|\ell''m''\rangle \, \langle \ell''m''|\ell_{q-\alpha}|\ell m\rangle \quad . \tag{d}$$

Using the Wigner-Eckert theorem on the first matrix element in (d) gives

$$\langle \ell m'|\ell_{\alpha}|\ell''m''\rangle = \langle \ell''(m'')1(\alpha)|\ell(m')\rangle \, \langle \ell''\|\ell\|\ell\rangle \quad . \tag{e}$$

44

But, as we have shown earlier, $\langle\ell''\|\ell\|\ell\rangle = \sqrt{\ell(\ell+1)}\ \delta_{\ell''\ell}$ and the C-G coefficient requires $m'' = m' - \alpha$. Substituting these results into (d) gives

$$\langle\ell m'|\ell_\alpha\ell_{q-\alpha}|\ell m\rangle = \langle\ell(m'-\alpha)1(\alpha)|\ell(m')\rangle\ \langle\ell(m)1(q-\alpha)|\ell(m'-\alpha)\rangle\ \ell(\ell+1) \qquad (f)$$

where we have used the same technique used in (e) on the second matrix element in (d).

The two C-G coefficients in (f) can be recoupled using equation (5.8) in the text to give

$$\langle\ell(m)1(q-\alpha)|\ell(m'-\alpha)\rangle\ \langle\ell(m'-\alpha)1(\alpha)|\ell(m')\rangle \qquad (g)$$

$$= \sum_f \sqrt{(2f+1)(2\ell+1)}\ W(\ell 1\ell 1;\ell f)\ \langle 1(q-\alpha)1(\alpha)|f(q)\rangle\ \langle\ell(m)f(q)|\ell(m')\rangle\ \ .$$

We now have the C-G coefficients uncoupled so that if we consider the sum on α given in (a) and the C-G coefficient in (g), we have

$$\sum_\alpha \langle 1(\alpha)1(q-\alpha)|k(q)\rangle\ \langle 1(q-\alpha)1(\alpha)|f(q)\rangle = (-1)^k\delta_{fk}$$

where we have used

$$\langle 1(\alpha)1(q-\alpha)|k(q)\rangle = (-1)^k\ \langle 1(q-\alpha)1(\alpha)|k(q)\rangle\ \ .$$

Collecting all these parts together--(g), (f), and (h)--and substituting into (c) we obtain

$$\langle\ell m'|T_{kq}|\ell m\rangle = (-1)^k\ell(\ell+1)[(2k+1)(2\ell+1)]^{1/2}W(\ell 1\ell 1;\ell k)\ \langle\ell(m)k(q)|\ell(m')\rangle\ \ .$$

Comparing this result to (b) we obtain

$$\langle\ell\|T_k\|\ell\rangle = (-1)^k\ell(\ell+1)[(2k+1)(2\ell+1)]^{1/2}\ W(\ell 1\ell 1;\ell k)\ \ . \qquad (h)$$

1. From the results obtained above, find the following:

for $k = 0$

$$\langle\ell\|T_0\|\ell\rangle = \ell(\ell+1)\ \sqrt{2\ell+1}\ W(\ell 1\ell 1;\ell 0)\ \ ,$$

$$W(\ell 1\ell 1;\ell 0) = \frac{-1}{\sqrt{3(2\ell+1)}} \qquad \text{(Rose, 1957, p 113)},$$

$$\langle\ell\|T_0\|\ell\rangle = \frac{-\ell(\ell+1)}{\sqrt{3}}\ \ .$$

But, from (a)

$$T_0 = \sum_\alpha \langle 1(\alpha)1(-\alpha)|0(0)\rangle \, \ell_\alpha \ell_{-\alpha}$$

and

$$\langle 1(\alpha)1(-\alpha)|0(0)\rangle = (-1)^{1-\alpha} \frac{1}{\sqrt{3}} \langle 1(\alpha)0(0)|1(\alpha)\rangle$$

$$= \frac{(-1)^{1-\alpha}}{\sqrt{3}} \quad .$$

Substituting the above,

$$T_0 = -\frac{1}{\sqrt{3}} \sum_\alpha (-1)^\alpha \ell_{-\alpha} \ell_\alpha$$

$$= -\frac{1}{\sqrt{3}} (\vec{\ell})^2 \quad ;$$

also,

$$(\vec{\ell})^2 |\ell m\rangle = \ell(\ell+1)|\ell m\rangle \quad ,$$

$$\langle \ell m'|T_0|\ell m\rangle = \langle \ell(m)0(0)|\ell(m)\rangle \, \langle \ell \| T_0 \| \ell\rangle$$

$$= -\frac{1}{\sqrt{3}} \ell(\ell+1) \quad .$$

Show also

$$W(\ell 1 \ell 1; \ell 1) = \frac{1}{\sqrt{6}\,[\ell(\ell+1)(2\ell+1)]^{1/2}} \quad ,$$

$$W(\ell 1 \ell 1; \ell 2) = \frac{\sqrt{(2\ell+3)(2\ell-1)}}{3\sqrt{30}\,\sqrt{\ell(\ell+1)(2\ell+1)}} \quad ,$$

$$\langle \ell \| T_1 \| \ell\rangle = -\frac{\sqrt{\ell(\ell+1)}}{\sqrt{2}} \quad , \quad \langle \ell \| T_2 \| \ell\rangle = \frac{[\ell(\ell+1)(2\ell+3)(2\ell-1)]^{1/2}}{3\sqrt{6}} \quad .$$

2. Given the results of problem 1, calculate the reduced matrix elements of the tensor

$$W_{KQ} = \sum_\beta \langle 1(\beta)k(Q-\beta)|K(Q)\rangle \, \ell_\beta T_{k,Q-\beta}$$

46

where T_{kq} is given in the discussion at the beginning of section 5.1.

The reduced matrix elements of W_{KQ} are given by the Wigner-Eckart theorem as

$$\langle \ell m' | W_{KQ} | \ell m \rangle = \langle \ell(m)K(Q) | \ell(m') \rangle \langle \ell \| W_K \| \ell \rangle \quad .$$

Then, by computing the matrix elements of W_{KQ} by the methods of problem 1, show that

$$\langle \ell \| W_K \| \ell \rangle = -(-1)^{k-K} [\ell(\ell+1)(2\ell+1)(2K+1)]^{1/2} \langle \ell \| T_k \| \ell \rangle W(\ell k \ell 1; \ell K) \quad .$$

Evaluate $\langle \ell \| W_k \| \ell \rangle$ for $K = 0, 1, 2, 3$. (The latter Racah coefficient is given in the appendix in Rose, 1957.)

Show the following:

$$W(\ell K \ell 1; \ell K) = \left[\frac{K(K+1)}{4K(K+1)(2K+1)\ell(\ell+1)(2\ell+1)} \right]^{1/2} \quad , \qquad k = K \quad ,$$

$$W(\ell K+1 \ell 1; \ell L) = - \left[\frac{(K+2\ell+2)(K+1)^2(2\ell-K)}{4(2K+1)(K+1)\ell(\ell+1)(2\ell+1)} \right]^{1/2} \quad , \qquad k = K + 1 \quad ,$$

$$W(\ell K-1 \ell 1; \ell K) = \left[\frac{(K+2\ell+1)K^2(2\ell+1-K)}{4(2K+1)K(2K-1)\ell(2\ell+1)(\ell+1)} \right]^{1/2} \quad , \qquad k = K - 1 \quad .$$

We write

$$\langle \ell \| W_K \| \ell \rangle = \langle \ell \| W_K(k) \| \ell \rangle \quad .$$

Then, when $K = 0$, we have $k = 1$ and

$$\langle \ell \| W_0(1) \| \ell \rangle = \frac{\ell(\ell+1)}{\sqrt{6}} \quad .$$

For $K = 1$, we have $k = 0, 1, 2,$ and

$$\langle \ell \| W_2(1) \| \ell \rangle = - \frac{[\ell(\ell+1)(2\ell-1)(2\ell+3)]^{1/2}}{2\sqrt{3}}$$

$$\langle \ell \| W_2(2) \| \ell \rangle = - \frac{[\ell(\ell+1)(2\ell-1)(2\ell+3)]^{1/2}}{6} \quad .$$

47

For K = 3, we have k = 2 and

$$\langle \ell \| W_3(2) \| \ell \rangle = \frac{[(\ell-1)\ell(\ell+1)(\ell+2)(2\ell-1)(2\ell+3)]^{1/2}}{3\sqrt{10}} \quad .$$

5.2 Bibliography and References

Brink, D. M., and G. R. Satchler (1962), Angular Momentum, Clarendon Press, Oxford, U.K., pp 40-45 and 116-118.

Edmonds, A. R. (1957), Angular Momentum in Quantum Mechanics, Princeton University Press, Princeton, NJ, pp 92-100.

Rose, M. E. (1957), Elementary Theory of Angular Momentum, Wiley, New York, NY, chapter VI, also pp 225-227.

Rotenberg, M., R. Bevins, N. Metropolis, and J. K. Wooten, Jr. (1969), The 3-j and 6-j Symbols, MIT Press, Cambridge, MA.

Sobelman, I. I. (1979), Atomic Spectra and Radiative Transitions, Springer-Verlag, New York, NY, pp 66-72.

6. RACAH ALGEBRA

It is convenient in many vector problems to express the vectors in terms of spherical bases given by

$$\hat{e}_{\pm 1} = \mp(\hat{e}_x \pm i\hat{e}_y)/\sqrt{2} \quad,$$

$$(6.1)$$

$$\hat{e}_0 = \hat{e}_z \quad.$$

Then

$$\hat{e}_\mu^* = (-1)^\mu \hat{e}_{-\mu} \quad,$$

$$\hat{e}_\mu \times \hat{e}_\nu = -i\sqrt{2} \langle 1(\nu)1(\mu)|1(\mu+\nu)\rangle \, \hat{e}_{\mu+\nu} \quad,$$

$$(6.2)$$

$$\hat{e}_\mu^* \cdot \hat{e}_\nu = \delta_{\mu\nu} \quad.$$

The vector $\vec{A}$ can be written

$$\vec{A} = \sum_\mu \hat{e}_\mu^* A_\mu \quad,$$

$$\vec{A} = \sum_\mu \hat{e}_\mu A_\mu^* \quad,$$

$$(6.3)$$

$$A_\mu = \hat{e}_\mu \cdot \vec{A} \quad,$$

and

$$\vec{A} \cdot \vec{B} = \sum_\mu A_\mu^* B_\mu$$

$$(6.4)$$

$$= \sum_\mu A_\mu B_\mu^*$$

$$= \sum_\mu (-1)^\mu A_{-\mu} B_\mu \quad.$$

Thus, $\vec{\ell} \cdot \vec{s}$ in equation (1.16) can be written

$$\vec{\ell} \cdot \vec{s} = \sum_\mu \ell^*_\mu s_\mu \tag{6.5}$$

so that the spin-orbit interaction given in equation (1.16) is immediately in spherical tensors, since ℓ_μ and s_μ are spherical tensors. That is, for the many-electron configuration,

$$H_{s-o} = \zeta(r) \sum_\mu \sqrt{\ell(\ell+1)s(s+1)} \, (-1)^\mu \, V^{11}_{-\mu\mu} \quad . \tag{6.6}$$

An example of Racah algebra is the reduction of the latter part of equation (4.6). That is, we wish to cast the interaction

$$\vec{I} \cdot \vec{T} = \vec{I} \cdot [-\vec{s} + 3\hat{r}(\hat{r} \cdot \vec{s})] \quad ,$$
$$\vec{I} \cdot \vec{T} = \sum_q I^*_q T_q \quad , \tag{6.7}$$

where we have dropped the subscript i on the components and introduced the unit vectors $\hat{r} = \vec{r}/r$. First we note that

$$\hat{r} = \sum_\alpha \hat{e}^*_\alpha C_{1\alpha}(\hat{r}) \tag{6.8}$$

as in equation (6.3), and we can write

$$\hat{r} \cdot \vec{s} = \sum_\beta (-1)^\beta C_{1-\beta}(\hat{r}) s_\beta \quad . \tag{6.9}$$

Then from (6.7) and (6.8)

$$T_q = -s_q + 3C_{1q}(\hat{r}) \sum_\beta (-1)^\beta C_{1-\beta}(\hat{r}) s_\beta \quad . \tag{6.10}$$

The recoupling given in equation (1.27) can be used to get

$$C_{1q}C_{1-\beta} = \sum_k \langle 1(0)1(0)|k(0)\rangle \, \langle 1(q)1(-\beta)|k(q-\beta)\rangle \, C_{k,q-\beta} \quad , \tag{6.11}$$

and since $\langle 1(0)1(0)|1(0)\rangle = 0$, the terms in (6.11) are restricted to $k = 0$ and $k = 2$. Then we have

$$C_{1q}C_{1-\beta} = \langle 1(0)1(0)|0(0)\rangle \langle 1(q)1(-q)|0(0)\rangle \delta_{\beta q}$$

$$+ \langle 1(0)1(0)|2(0)\rangle \langle 1(q)1(-\beta)|2(q-\beta)\rangle C_{2,q-\beta} \quad . \tag{6.12}$$

From the relation equation (2.10) we have $\langle 1(q)0(0)|1(q)\rangle = 1$ and by symmetry (eq (2.7))

$$\langle 1(q)1(-q)|0(0)\rangle = (-1)^{1-q}/\sqrt{3} \tag{6.13}$$

and

$$C_{1q}C_{1-\beta} = \frac{(-1)^q}{3} \delta_{\beta,q} - \frac{\sqrt{10}}{3} (-1)^\beta \langle 1(\beta)2(q-\beta)|1(q)\rangle C_{2,q-\beta} \quad , \tag{6.14}$$

where we have used the symmetry relation of equation (2.7) on the C-G coefficient $\langle 1(q)1(-\beta)|2(q-\beta)\rangle$; also $\langle 1(0)1(0)|2(0)\rangle = -(2/5)^{1/2}$. The first term in equation (6.14), when substituted into equation (6.10), cancels the $-s_q$ term, and the remainder gives

$$T_q = - \sqrt{10} \sum_\beta \langle 1(\beta)2(q-\beta)|1(q)\rangle s_\beta C_{2,q-\beta} \quad , \tag{6.15}$$

which is the form used in equation (4.7) where this interaction was cast into V^{12} tensors.

As a further application of Racah algebra and some of the other material discussed above, we shall derive the gradient formula (Rose, 1957, p 120). A convenient form of the gradient operator is

$$\nabla = \hat{r} \frac{\partial}{\partial r} - i \frac{\hat{r} \times \hat{\ell}}{r} \quad , \tag{6.16}$$

and we would like

$$\text{grad } \phi(r)C_{kq}(\hat{r}) = [\nabla, \phi(r)C_{kq}] \quad . \tag{6.17}$$

First we observe that

$$rC_{kq} = \sum_\lambda (-1)^\lambda \hat{e}_{-\lambda} C_{1\lambda} C_{kq}$$

$$= \sum_{\lambda,k'} (-1)^\lambda \hat{e}_{-\lambda} \langle 1(0)k(0)|k'(0)\rangle \langle 1(\lambda)k(q)|k'(q+\lambda)\rangle C_{k',q+\lambda} \quad , \tag{6.18}$$

where we have used the coupling rule for spherical harmonics, equation (1.27) (Rose, 1957, p 61). Now we write

$$\hat{r} \times \vec{\ell} = \sum_{\alpha\beta} (-1)^{\alpha+\beta} \hat{e}_\alpha \times \hat{e}_\beta C_{1-\alpha} \ell_{-\beta} \quad , \tag{6.19}$$

and we use equation (6.2) to eliminate the cross product to produce

$$\hat{r} \times \vec{\ell} = -i\sqrt{2} \sum_{\lambda,\alpha} (-1)^\lambda \langle 1(\lambda-\alpha)1(\alpha)|1(\lambda)\rangle \hat{e}_\lambda C_{1-\alpha} \ell_{\alpha-\lambda} \quad , \tag{6.20}$$

where we have replaced the sum on β by letting $\beta = \lambda - \alpha$. Now in calculating the commutation we need only consider the operators in equation (6.20); thus, we need

$$[C_{1-\alpha} \ell_{\alpha-\lambda}, C_{kq}] \quad . \tag{6.21}$$

Since $\phi(r)$ commutes with $C_{1-\alpha} \ell_{\lambda-\alpha'}$ we need not consider it at present. First we expand the commutator to obtain

$$[C_{1-\alpha} \ell_{\alpha-\lambda}, C_{kq}] = C_{1-\alpha} \ell_{\alpha-\lambda} C_{kq} - C_{kq} C_{1-\alpha} \ell_{\alpha-\lambda} \quad ; \tag{6.22}$$

we then use

$$\ell_{\alpha-\lambda} C_{kq} = [\ell_{\alpha-\lambda}, C_{kq}] + C_{kq} \ell_{\alpha-\lambda} \tag{6.23}$$

in equation (6.22) to obtain

$$[C_{1-\alpha} \ell_{\alpha-\lambda}, C_{kq}] = C_{1-\alpha}[\ell_{\alpha-\lambda}, C_{kq}] + C_{1-\alpha} C_{kq} \ell_{\alpha-\lambda} - C_{kq} C_{1-\alpha} \ell_{\alpha-\lambda} \quad . \tag{6.24}$$

The last two terms cancel since $C_{1-\alpha}$ and C_{kq} commute. Thus, we obtain

$$[C_{1-\alpha}\ell_{\alpha-\lambda},C_{kq}] = C_{1-\alpha}[\ell_{\alpha-\lambda},C_{kq}] \tag{6.25}$$

$$[C_{1-\alpha}\ell_{\alpha-\lambda},C_{kq}] = C_{1-\alpha}\sqrt{k(k+1)}\ \langle k(q)1(\alpha-\lambda)|k(q+\alpha-\lambda)\rangle\ C_{k,q+\alpha-\lambda}\ , \tag{6.26}$$

where we have used equation (2.19) with $L_{\alpha-\lambda} = \ell_{\alpha-\lambda}$ (which are identical in the commutation brackets). The result in equation (6.26) is not quite in the form we want, but by using the coupling rule for spherical harmonics given in equation (1.27), we finally obtain

$$[C_{1-\alpha}\ell_{\alpha-\lambda},C_{kq}] = \sqrt{k(k+1)}\ \langle k(q)1(\alpha-\lambda)|k(q+\alpha-\lambda)\rangle$$

$$\tag{6.27}$$

$$\times \sum_{k''}\ \langle k(0)1(0)|k''(0)\rangle\ \langle k(q-\alpha-\lambda)1(-\alpha)|k''(q-\lambda)\rangle\ C_{k'',q-\lambda}\ .$$

In equations (6.16), (6.17), and (6.20), we need

$$[\vec{r}\times\vec{\ell},C_{kq}]\ . \tag{6.28}$$

We can see from equations (6.27) and (6.20) that, when this is formed, the terms dependent on α are

$$S = \sum_{\alpha}\ \langle 1(\lambda-\alpha)1(\alpha)|1(\lambda)\rangle\ \langle k(q)1(\alpha-\lambda)|k(q+\alpha-\lambda)\rangle$$

$$\tag{6.29}$$

$$\times\ \langle k(q+\lambda-\alpha)1(-\alpha)|k''(q-\lambda)\rangle\ ;$$

that is,

$$[\vec{r}\times\vec{\ell},C_{kq}] = i\ \sqrt{2}\ \sum_{\lambda\alpha m k''}\ (-1)^{\lambda}\hat{e}_{\lambda}\ \langle k(0)1(0)|k''(0)\rangle\ \sqrt{k(k+1)}\ SC_{k'',q-\lambda} \tag{6.30}$$

with S given by equation (6.29). The sum, S, given by equation (6.29) can be reduced. First we write

$$\langle k(q)1(\alpha-\lambda)|k(q+\alpha-\lambda)\rangle\ \langle k(q+\alpha-\lambda)1(-\alpha)|k''(q-\lambda)\rangle$$

$$\tag{6.31}$$

$$= \sum_{f}\ \sqrt{(2f+1)(2k+1)}\ W(k1k''1;kf)\ \langle 1(\alpha-\lambda)1(-\alpha)|f(-\lambda)\rangle\ \langle k(q)f(-\lambda)|k''(q-\lambda)\rangle\ ,$$

where we have used equation (5.8). Thus, the sum over α contains the terms

$$\sum_\alpha \langle 1(\alpha-\lambda)1(-\alpha)|1(-\lambda)\rangle \langle 1(\alpha-\lambda)1(-\alpha)|f(-\lambda)\rangle = \delta_{f1} \qquad (6.32)$$

by the orthogonality of the C-G coefficients. We can use equations (6.32) and (6.31) in equation (6.29) to obtain

$$S = \sqrt{3(2k+1)}\ W(k1k''1;k1)\ \langle k(q)1(-\lambda)|k''(q-\lambda)\rangle \quad . \qquad (6.33)$$

Using the results of equation (6.33) in equation (6.30) gives

$$[\vec{r}\times\vec{\ell},C_{kq}] = i\sqrt{6k(k+1)(2k+1)} \sum_\lambda (-1)^\lambda \hat{e}_{-\lambda} \sum_{k''} \langle k(0)1(0)|k(0)\rangle\ W\ \langle k1k''1;k1\rangle \qquad (6.34)$$

$$\times\ \langle k(q)1(\lambda)|k''(q+\lambda)\rangle\ C_{k'',q+\lambda} \quad ,$$

where we have changed the sign of λ in the sum. Multiplying the results given in equation (6.34) by $-i\phi(r)/r$ and combining them with equation (6.28), we have (changing k'' to k')

$$[\nabla,\phi(r)C_{kq}] = \sum_\lambda (-1)^\lambda \hat{e}_{-\lambda} \sum_{k'} \left[\frac{\partial\phi}{\partial r} + \frac{\phi}{r}\sqrt{6k(k+1)(2k+1)}\ W(k1k'1;k1)\right] \qquad (6.35)$$

$$\times\ \langle k(0)1(0)|k'(0)\rangle\ \langle k(q)1(\lambda)|k'(q+\lambda)\rangle\ C_{k',q+\lambda} \quad .$$

The Racah coefficients in equation (6.35) are of simple form and are given by Rose (1957, p 227). These are

$$W(k1k'1;k1) = -\left[\frac{k}{6(k+1)(2k+1)}\right]^{1/2} \quad , \qquad\qquad k' = k + 1 \quad ,$$

$$(6.36)$$

$$= \left[\frac{k+1}{6k(2k+1)}\right]^{1/2} \quad , \qquad\qquad k' = k - 1 \quad ,$$

which are the only values of k' allowed. These results used in equation (6.18) can be written as

$$[\nabla,\phi(r)C_{kq}] = \sum_\lambda (-1)^\lambda \hat{e}_{-\lambda} \sum_{k'} \langle k(q)1(\lambda)|k'(q+\lambda)\rangle\ C_{k',q+\lambda}D^{k'}\phi(r) \quad ,$$

$$(6.37)$$

where

$$D^{k'} = \left[\frac{k+1}{2k+1}\right]^{1/2}\left(\frac{\partial}{\partial r} - \frac{k}{r}\right) \quad , \qquad k' = k + 1 \quad , \qquad (6.37a)$$

$$D^{k'} = -\left[\frac{k}{2k+1}\right]^{1/2}\left(\frac{\partial}{\partial r} + \frac{k+1}{r}\right) \quad , \qquad k' = k - 1 \quad , \qquad (6.37b)$$

and we have used the result

$$<k(0)1(0)|k+1(0)> = \left[\frac{k+1}{2k+1}\right]^{1/2} \quad \text{and}$$

$$<k(0)1(0)|k-1(0)> = -\left[\frac{k}{2k+1}\right]^{1/2}$$

(from Rose, 1957, p 225). The two most common forms of $\phi(r)$ that we will encounter are r^k and $1/r^{k+1}$. For $\phi(r) = r^k$, we obtain

$$\left[\nabla_\mu, r^k C_{kq}\right] = -\sqrt{k(2k+1)} <k(q)1(\mu)|k-1(q+\mu)> r^{k-1}C_{k-1,q+\mu} \quad , \qquad (6.38)$$

and for $\phi = 1/r^{k+1}$,

$$\left[\nabla_\mu, 1/r^{k+1} C_{kq}\right] = -\sqrt{(k+1)(2k+1)} <k(q)1(\mu)|k+1(q+\mu)> 1/r^{k+2}C_{k+1,q+\mu} \quad .$$
$$(6.39)$$

The results given in equation (6.39) are easily checked for k = 0, since for k = 0 we have

$$\left[\nabla_\mu, 1/r\right] = -<0(q)1(\mu)|1(q+\mu)> 1/r^2 C_{1,q+\mu} \quad , \qquad (6.39a)$$

and from the properties of the C-G coefficients, we know that q = 0 and $<0(0)1(\mu)|1(\mu)> = 1$. Then

$$\left[\nabla_\mu, 1/r C_{1q}\right] = -C_{1\mu}/r^2 \quad . \qquad (6.39b)$$

Also, we know from vector analysis that

$$\text{grad } 1/r = -\vec{r}/r^3 \qquad (6.39c)$$

55

and

$$\vec{r} = r \sum_\mu (-1)^\mu \hat{e}_{-\mu} C_{1\mu} \quad . \tag{6.39d}$$

Then we substitute equation (6.39d) in (6.39c) to obtain

$$(\text{grad } 1/r)_\mu = -C_{1\mu}/r^2 \quad , \tag{6.39e}$$

which is identical with the result of equation (6.39b). We shall use the result given in equation (6.35) frequently later on, particularly in the form given in equations (6.38) and (6.39).

6.1 Problems

1. In section 3.1 we considered the crystal-field Hamiltonian in the form $H_{CEF} = B_{20}C_{20}(\hat{r}) + B_{40}C_{40}(\hat{r})$ and obtained the energy splittings of a single d electron. All the previous calculation was for the orbital states only, neglecting the spin-orbit coupling. For the states $|jm\rangle$ with $\vec{j} = \vec{\ell} + \vec{s}$ we showed (in sect. 2.1, Problems) that

$$\langle j'm'\ell s|H_{s-o}|jm\ell s\rangle = \delta_{j'j}\delta_{m'm} \frac{\zeta}{2} [j(j+1) - \ell(\ell+1) - s(s+1)] \quad ,$$

where the states $|jm\ell s\rangle$ are

$$|jm\ell s\rangle = \sum_\mu \langle \ell(\mu)s(m-\mu)|j(m)\rangle \; |\ell\mu\rangle|sm-\mu\rangle \quad .$$

For these same states show that

$$\langle j'\ell s|C_k|j\ell s\rangle = (-1)^{j'-j}\sqrt{(2j+1)(2\ell+1)} \; W(k\ell j's;\ell j) \; \langle \ell(0)k(0)|\ell(0)\rangle \quad ,$$

and evaluate this quantity for

$$\ell = 2 \quad ,$$

$$s = \frac{1}{2} \quad ,$$

$$j \text{ or } j' = \frac{3}{2}, \; \frac{5}{2} \text{ (for } \ell-s \leq j \leq \ell+s),$$

$$k = 2, 4 \quad ,$$

$$W(k\ell j's;\ell j) = (-1)^{k+\ell+j'+s} \begin{Bmatrix} k & \ell & \ell \\ s & j' & j' \end{Bmatrix} \quad ,$$

and the quantity { } is a 6-j symbol which can be found in Rotenberg et al (1969). Using the above results calculate

$$\langle j'm|C_{k0}|jm\rangle = \langle j(m)k(0)|j'(m)\rangle \langle j'\ell s|C_k|j\ell s\rangle$$

and obtain the following table for the matrix elements of the crystal field.

j'	m'	j	m	B_{20}	B_{40}
$\frac{3}{2}$	$\frac{1}{2}$	$\frac{3}{2}$	$\frac{1}{2}$	$1/5$	0
$\frac{5}{2}$	$\frac{1}{2}$	$\frac{5}{2}$	$\frac{1}{2}$	$8/35$	$2/21$
$\frac{3}{2}$	$\frac{1}{2}$	$\frac{5}{2}$	$\frac{1}{2}$	$-\sqrt{6}/35$	$-2\sqrt{6}/21$
$\frac{3}{2}$	$\frac{3}{2}$	$\frac{3}{2}$	$\frac{3}{2}$	$-1/5$	0
$\frac{5}{2}$	$\frac{3}{2}$	$\frac{5}{2}$	$\frac{3}{2}$	$2/35$	$-1/7$
$\frac{5}{2}$	$\frac{5}{2}$	$\frac{5}{2}$	$\frac{5}{2}$	$-2/7$	$1/21$
$\frac{3}{2}$	$\frac{3}{2}$	$\frac{5}{2}$	$\frac{3}{2}$	$-6/35$	$2/21$

To obtain the energy levels we let

$$\langle \tfrac{3}{2}\ \tfrac{1}{2}|H_{CEF}|\tfrac{3}{2}\ \tfrac{1}{2}\rangle = H_{11}\ ,$$

$$\langle \tfrac{5}{2}\ \tfrac{1}{2}|H_{CEF}|\tfrac{5}{2}\ \tfrac{1}{2}\rangle = H_{22}\ ,\ \text{and}$$

$$\langle \tfrac{3}{2}\ \tfrac{1}{2}|H_{CEF}|\tfrac{5}{2}\ \tfrac{1}{2}\rangle = H_{12}$$

$(H_{21} = H_{12})$. Then, for the $m = \frac{1}{2}$ levels, we have the secular equation

$$\begin{vmatrix} H_{11} + H_{s-o}\left(j = \tfrac{3}{2}\right) - E & H_{12} \\ H_{12} & H_{22} + H_{s-o}\left(j = \tfrac{5}{2}\right) - E \end{vmatrix} = 0$$

57

with a similar result for $m = \frac{3}{2}$. What is the result for $m = \frac{5}{2}$?

2. In the consideration of the nonrelativistic limit of the Dirac equation, a correction to the nonrelativistic Hamiltonian arises of the form

$$H' = -\frac{\alpha^2 a_o^2}{4}\left[\frac{\partial V}{\partial r}\frac{\partial}{\partial r} - \frac{1}{r^2}\,\vec{T}\cdot\vec{\ell}\right] \tag{a}$$

for terms involving the orbital motion only. In equation (a),

α = the fine structure constant = $\frac{e^2}{\hbar c}$,

a_o = the first Bohr radius = $\frac{\hbar^2}{me^2}$,

V = the potential energy (not necessarily spherically symmetric), $\vec{T} = [\vec{\ell}, V]$.

Take $V = f_k(r)C_{kq}(\hat{r})$ and show that

$$\vec{T}\cdot\vec{\ell} = \sum_\lambda (-1)^\lambda\,[\ell_\lambda, C_{kq}]\ell_{-\lambda}f_k(r)$$

$$= f_k(r)\,\sqrt{k(k+1)}\,\sum_\lambda (-1)^\lambda\,\langle k(q)1(\lambda)|k(q+\lambda)\rangle\,C_{kq+\lambda}\ell_{-\lambda}\ .$$

Using this result show that

$$H' = -\alpha\frac{2a_o^2}{4}\left[\frac{\partial f_k(r)}{\partial r}\frac{\partial}{\partial r}\,C_{kq}(\hat{r}) + \frac{f_k(r)}{r^2}\,\sqrt{k(k+1)}\,\sum_\lambda \langle 1(-\lambda)k(q+\lambda)|k(q)\rangle\,C_{kq+\lambda}\ell_{-\lambda}\right]\ .$$

Further, show that

$$\langle \ell'm'|\sum_\lambda \langle 1(-\lambda)k(q+\lambda)|k(q)\rangle\,C_{kq+\lambda}\ell_{-\lambda}|\ell m\rangle$$

$$= [\ell(\ell+1)(2\ell+1)(2k+1)]^{1/2}W(\ell 1\ell'k;\ell k)\,\langle \ell(m)k(q)|\ell'(m')\rangle\,\langle \ell'|C_k|\ell\rangle$$

and

$$\langle \ell'm'|H'|\ell\rangle = -\frac{\alpha^2 a_o^2}{4} F_{k\ell\ell'}(r) \langle \ell(m)k(q)|\ell'(m')\rangle \langle \ell'|C_k|\ell\rangle \quad,$$

$$F_{k\ell\ell'}(r) = \frac{\partial f_k}{\partial r}\frac{\partial}{\partial r} + \frac{f_k(r)}{r^2}[\ell(\ell+1)(2\ell+1)k(k+1)(2k+1)]^{1/2}W(\ell 1\ell'k;\ell k) \quad,$$

$$W(\ell 1\ell'k;\ell k) = \frac{\ell(\ell+1) + k(k+1) - \ell'(\ell'+1)}{2[\ell(\ell+1)(2\ell+1)k(k+1)(2k+1)]^{1/2}} \quad,$$

$$F_{k\ell\ell'}(r) = \frac{\partial f_k}{\partial r}\frac{\partial}{\partial r} + \frac{f_k}{2r^2}[\ell(\ell+1) + k(k+1) - \ell'(\ell'+1)] \quad.$$

If $\ell' = \ell$ and $f_k = C_o r^k$ (C_o is a constant), show that

$$\int r^2\, dr\, R_\ell F_{k\ell\ell}(r)R_\ell = 0 \quad,$$

that is,

$$\langle kr^{k-1}\frac{\partial}{\partial r}\rangle = -\frac{k(k+1)}{2}\langle r^{k-2}\rangle \quad.$$

3. In the basis of states given by

$$|jm\rangle = \sum_\mu \langle \ell(\mu)s(m-\mu)|j(m)\rangle\, |\ell\mu\rangle|sm-\mu\rangle \quad,$$

calculate the matrix elements of ℓ_λ and the matrix elements of s_λ. Show that the following statements are true:

$$\langle j'\|\ell\|j\rangle = (-1)^{j'-j}[\ell(\ell+1)(2\ell+1)(2j+1)]^{1/2}W(1\ell j's;\ell j) \quad,$$

$$\langle j'\|s\|j\rangle = [s(s+1)(2s+1)(2j+1)]^{1/2}W(1sj'\ell;sj) \quad.$$

4. We write a generalized spin-orbit interaction as the mixed tensor

$$T_{kq} = \sum_\alpha \langle 1(\alpha)1(q-\alpha)|k(q)\rangle\, \ell_\alpha s_{q-\alpha}$$

and we wish to calculate matrix elements of T_{kq} in the coupled space (single electron) $|jm\rangle$ with $\vec{j} = \vec{\ell} + \vec{s}$. By the Wigner-Eckart theorem,

$$\langle j'm'|T_{kq}|jm\rangle = \langle j(m)k(q)|j'(m')\rangle \langle j\|T_k\|j\rangle \quad , \tag{a}$$

which is the easy part. We now write the state $|jm\rangle$ as

$$|jm\rangle = \sum_{\mu} \langle \ell(\mu)s(m-\mu)|j(m)\rangle \; |\ell\mu\rangle|sm-\mu\rangle \tag{b}$$

and a similar expression for $\langle j'm'|$, so that the matrix elements given in (a) are

$$\langle j'm'|T_{kq}|jm\rangle = \sum_{\alpha} \langle 1(\alpha)1(q-\alpha)|k(q)\rangle \sum_{\mu\mu'} \langle \ell(\mu)s(m-\mu)|j(m)\rangle$$

$$\times \langle \ell(\mu')s(m'-\mu')|j'(m')\rangle \langle \ell\mu'|\ell_{\alpha}|\ell\mu\rangle \tag{c}$$

$$\times \langle sm'-\mu'|s_{q-\alpha}|sm-\mu\rangle \quad .$$

The Wigner-Eckart theorem can be used on the last two matrix elements in (c) to give

$$\langle \ell\mu'|\ell_{\alpha}|\ell\mu\rangle = \langle \ell(\mu)1(\alpha)|\ell(\mu')\rangle \sqrt{\ell(\ell+1)}$$

$$\tag{d}$$

$$\langle sm'-\mu'|s_{q-\alpha}|sm-\mu\rangle = \langle s(m-\mu)1(q-\alpha)|s(m'-\mu')\rangle \sqrt{s(s+1)}$$

where we have used $\langle \ell\|\ell\|\ell\rangle = \sqrt{\ell(\ell+1)}$ and $\langle s\|s\|s\rangle = \sqrt{s(s+1)}$.

The first C-G coefficient in (d) requires $\mu'+\mu+\alpha$, and the second C-G requires $m-\mu+q-\alpha m'-\mu'$; when μ' is used, this results in $m' = m+q$, which agrees with the restriction on m' given by the C-G coefficient in (a). Substituting the results of (d) into (c) gives

$$\overset{1}{\overbrace{}} \qquad \overset{2}{\overbrace{}} \qquad \overset{3}{\overbrace{}}$$

$$\langle j'm'|T_{kq}|jm\rangle = \sum_{\mu\alpha} \langle 1(\alpha)1(q-\alpha)|k(q)\rangle \; \langle \ell(\mu)s(m-\mu)|j(m)\rangle \; \langle \ell(\mu+\alpha)s(m+q+\mu-\alpha)|j'(m')\rangle$$

$$\times \underset{4}{\underbrace{\langle \ell(\mu)1(\alpha)|\ell(\mu+\alpha)\rangle}} \; \underset{5}{\underbrace{\langle s(m-\mu)1(\mu-\alpha)|s(m+q-\mu-\alpha)\rangle}} \; [s(s+1)\ell(\ell+1)]^{1/2}$$

$$\tag{e}$$

60

and we shall refer to the C-G coefficients by a number referring to the position in the sum in equation (e). First we recouple 4 and 3 using equation (5.9) to give

$$\langle\ell(\mu)1(\alpha)|\ell(\mu+\alpha)\rangle \ \langle\ell(\mu+\alpha)s(m+q-\mu-\alpha)|j'(m')\rangle$$

$$= (-1) \sum_f \sqrt{(2f+1)(2\ell+1)} \ W(1\ell j's;\ell f) \ \langle\ell(\mu)s(m+q-\mu-\alpha)|f(m+q-\alpha)\rangle$$

$$\text{(f)}$$

$$\times \ \langle1(\alpha)f(m+q-\alpha)|j'(m')\rangle$$

where we have used the symmetry of the C-G on 4 to reverse the order of the first two angular momenta. We now recouple 5 in equation (e) with the first C-G in equation (f) to give

$$\langle s(m-\mu)1(q-\alpha)|s(m+q-\mu-\alpha)\rangle \ \langle\ell(\mu)s(m+q-\mu-\alpha)|f(m+q-\alpha)\rangle$$

$$= (-1)^{\ell+s-f+1} \sum_g \sqrt{(2q+1)(2s+1)} \ W(1sf\ell;sg) \qquad\qquad \text{(g)}$$

$$\times \ \langle s(m-\mu)\ell(\mu)|g(m)\rangle \ \langle1(q-\alpha)g(m)|f(m+q-\alpha)\rangle$$

where we have used the symmetry of the C-G to rearrange the order of the angular momentum in both coefficients. The second C-G coefficient in (e) and the first C-G coefficient in (g) are the only coefficients containing μ, and the sum is

$$\sum_\mu \langle s(m-\mu)\ell(\mu)|g(m)\rangle \ \langle\ell(\mu)s(m-\mu)|j(m)\rangle = (-1)^{j-\ell-s}\delta_{gj} \ . \qquad \text{(h)}$$

Collecting what remains of (e) using the results of (f), (g), and (h), we have

$$\langle j'm'|T_{kq}|jm\rangle = (-1)^j[s(s+1)\ell(\ell+1)(2s+1)(2j+1)(2\ell+1)]^{1/2}$$

$$\times \sum_f (-1)^f\sqrt{2f+1} \ W(1\ell j's;\ell f)W(1sf\ell;sj)$$

$$\text{(e')}$$

$$\times \sum_\alpha \langle1(\alpha)1(q-\alpha)|k(q)\rangle \ \langle1(q-\alpha)j(m)|f(m+q-\alpha)\rangle$$

$$\times \ \langle1(\alpha)f(m+q-\alpha)|j'(m')\rangle \ .$$

We now recouple the last two C-G coefficients in (e') to obtain

$$\langle 1(q-\alpha)j(m)|f(m+q-\alpha)\rangle \langle 1(\alpha)f(m+q-\alpha)|j'(m')\rangle$$

$$= (-1)^{j'-j} \langle j(m)1(q-\alpha)|f(m+q-\alpha)\rangle \langle f(m+q-\alpha)1(\alpha)|j'(m')\rangle \qquad (i)$$

$$= (-1)^{j'-j} \sum_h \sqrt{(2h+1)(2f+1)}\; W(j1j'1;fh)\; \langle 1(q-\alpha)1(\alpha)|h(q)\rangle$$

$$\times \langle j(m)h(q)|j'(m')\rangle \quad .$$

Finally, the sum over α with the remaining C-G in (e') and the first C-G in (i) gives

$$\sum_\alpha \langle 1(\alpha)1(q-\alpha)|k(q)\rangle \langle 1(q-\alpha)1(\alpha)|h(q)\rangle = (-1)^k \delta_{hk} \quad ; \qquad (j)$$

substituting into (5a) produces

$$\langle j'm'|T_{kq}|jm\rangle = (-1)^{k+j'}[s(s+1)(2s+1)\ell(\ell+1)(2\ell+1)(2j+1)(2k+1)]^{1/2}$$

$$(e'')$$

$$\times \langle j(m)k(q)|j'm'\rangle\, S$$

where

$$S = \sum_f (-1)^f (2f+1)\, W(1\ell j's;\ell f)\, W(1sf\ell;sj)\, W(j1j'1;fk) \quad .$$

From Rose (1957, p 191) or from Judd (1963, p 64), we have

$$S = (-1)^j X(11k;\ell sj;\ell sj') \qquad (k)$$

where X is the X-coefficient and is identical to a 9-j symbol.

By comparing (e'') with equation (a) we obtain

$$\langle j'\|T_k\|j\rangle = (-1)^{k+j'+j} [s(s+1)(2s+1)\ell(\ell+1)(2\ell+1)(2j+1)(2k+1)]^{1/2}$$

$$(e''')$$

$$\times X(11k;\ell sj;\ell sj') \quad .$$

From the rules for rearranging the arguments of an X-coefficient, we have the above equal to

$$X(\ell sj;\ell sj';11k)$$

and when k = 0 we get (Judd, 1963, p 67; Rose, 1957, p 192)

$$X(\ell sj;\ell sj';110) = \frac{(-1)^{j+1-\ell-s}}{[s(2j+1)]^{1/2}} \, W(\ell s\ell s;j1) \; \delta_{j'j} \qquad (1)$$

(see also Brink and Satchler, 1962, p 119). The latter reference gives

$$X(\ell sj;\ell sj;111) = 0 \;\; ;$$

therefore

$$\langle j'|T_1|j\rangle = 0 \;\; .$$

From equations (1) and (e'''), we have

$$\langle j'|T_0|j\rangle = \frac{(-1)^{\ell+s+1+j}[s(s+1)(2s+1)\ell(\ell+1)(2\ell+1)]^{1/2}}{\sqrt{3}} \, W(\ell s\ell s;h1) \;\; . \qquad (m)$$

From Rose (1957, p 227), we have

$$W(\ell s\ell s;j1) = \frac{(-1)^{-\ell-s+j+1}[\ell(\ell+1) + s(s+1) - j(j+1)]}{2[\ell(\ell+1)(2\ell+1)s(s+1)(2s+1)]^{1/2}} \qquad (n)$$

and from equation (m)

$$\langle j'|T_0|j\rangle = \frac{(-1)^{2j}}{2\sqrt{3}} \, [\ell(\ell+1) + s(s+1) - j(j+1)] \;\; . \qquad (o)$$

Now when k = 0 in T_{kq} as defined at the beginning,

$$T_{00} = \sum_{\alpha} \langle 1(\alpha)1(q-\alpha)|0(0)\rangle \, \ell_{\alpha} s_{q-\alpha} \delta_{q,0}$$

$$= \sum_{\alpha} \frac{(-1)^{1-\alpha}}{\sqrt{3}} \langle 1(\alpha)0(0)|1(\alpha)\rangle \, \ell_{\alpha} s_{-\alpha} \qquad (p)$$

$$= -\frac{1}{\sqrt{3}} \sum_{\alpha} (-1)^{\alpha} \ell_{\alpha} s_{-\alpha}$$

$$= -\frac{1}{\sqrt{3}} \, \vec{\ell}\cdot\vec{s} \;\; .$$

In section 6 we have shown that

$$\vec{\ell} \cdot \vec{s} = 1/2 \ [\vec{j}^2 - \vec{\ell}^2 - \vec{s}^2] \quad ;$$

therefore

$$\langle j'm'|T_0|jm\rangle = - \frac{1}{2\sqrt{3}} \ [j(j+1) - \ell(\ell+1) - s(s+1)] \ \delta_{m'm}\delta_{j'j} \qquad (q)$$

and in equation (o), $j = \ell \pm s = \ell \pm 1/2$, $2j = 2\ell \pm 1$; therefore,

$$(-1)^{2j} = -1 \quad ,$$

so that the result given in equation (o) agrees with equation (q) but is much more difficult to obtain.

We have an interaction between two electrons given by

$$T_{kq} = \sum_\alpha \langle a(\alpha)b(q-\alpha)|k(q)\rangle \ C_{a\alpha}(1)C_{b,q-\alpha}(2) \quad .$$

By the above techniques obtain the reduced matrix element in the expression

$$\langle L'M'|T_{kq}|LM\rangle = \langle L(M)k(q)|L'(M')\rangle \ \langle L'\|T_k\|L\rangle$$

where the states are given by

$$|LM\rangle = \sum_\mu \langle \ell(\mu)\ell(M-\mu)|L(M)\rangle \ |\ell\mu;1\rangle|\ell M-\mu;2\rangle$$

where the last index in the angled brackets refers to the individual electrons. Hint: The result should have something like

$$X(abk;\ell\ell L;\ell\ell L) \quad .$$

6.2 Annotated Bibliography and References

Brink, D. M., and G. R. Satchler (1962), Angular Momentum, Clarendon Press, Oxford, U.K., pp 79 and 82.

Edmonds, A. R. (1957), Angular Momentum in Quantum Mechanics, Princeton University Press, Princeton, NJ, pp 78-85.

Jackson, J. D. (1975), Classical Electrodynamics, Wiley, New York, NY.

Judd, B. R. (1963), Operator Techniques in Atomic Spectroscopy, McGraw-Hill, New York, NY.

Rose, M. E. (1957), Elementary Theory of Angular Momentum, Wiley, New York, NY. Pages 103-106 discuss the unit vectors $\hat{e}\alpha$. Pages 120-124 discuss the derivation of the gradient formula. Chapter VII has many applications of Racah algebra in electromagnetic theory and multipole fields. Chapter VIII covers static multipolar interactions.

Rose, M. E. (1955), Multipole Fields, Wiley, New York, NY. Practically the entire book is applications of Racah algebra. The Clebsch-Gordan coefficients are expressed in terms of $_3F_2$, and the Racah coefficients are expressed in terms of $_4F_3$, where the $_pF_q$ are hypergeometric functions.

Rotenberg, M., R. Bevins, N. Metropolis, and J. K. Wooten, Jr. (1969), The 3-j and 6-j Symbols, MIT Press, Cambridge, MA.

7. FREE-ION HAMILTONIAN UNFILLED CONFIGURATION $n\ell^N$ [N < 2(2ℓ + 1)]

7.1 Background for Free Ions

      The approximations made in the analysis of the spectra of ions are not new. In fact, they go back to the old Bohr orbit theory. Since many readers may not be familiar with these assumptions and may not remember many of the concepts and most of the technical jargon used in the field of atomic spectra, we review some of these briefly. We stick strictly to those concepts which apply to transition-metal ions and rare-earth ions.

      For the transition-metal ions we shall consider the doubly, triply, and quadruply ionized states. For the rare-earth ions we consider only the triply ionized case.

      The electronic structure of the three series of doubly ionized transition-metal ions is given in table 7.1. The triply ionized rare-earth ions are characterized by the electronic structure shown in table 7.2. In the rare-earth series, it is assumed that the atomic interactions are very strong; thus, when an ion is placed in a crystal, the crystalline electric field acts as a perturbation on the ion. In the transition-metal ions, the electronic interaction with the crystal can in some cases be larger than the free-ion interaction. Nevertheless, we shall continue to label the states using the free-ion labels. This assumption allows the notation developed for the free ion to be used, with the reservation that many of the "good" quantum numbers of the free ion are not quite good when the ion is present in the crystal. It is assumed that the free ions have the zeroth-order Hamiltonian

$$H_o = \sum_{i=1}^{N} \left[\frac{p_i^2}{2m} + U(r_i) \right] \quad , \tag{7.1}$$

where $\vec{p}_i$ is the momentum of the ith electron and $U(r_i)$ is an appropriate spherical average potential of the remaining electrons in the ion (other than the N, nℓN). The single-electron solutions to equation (7.1) are taken in the form

$$\psi = R_{n\ell}(r)Y_{\ell m}(\hat{r}) \tag{7.2}$$

TABLE 7.1. ELECTRONIC STRUCTURE OF TRANSITION-METAL IONS

First series[a]			Second series[b]			Third series[c]		
Z	X^{2+}	$3d^N$	Z	X^{2+}	$4d^N$	Z	X^{2+}	$5d^N$
21	Sc	$3d^1$	39	Y	$4d^1$	71	Lu	$5d^1$
22	Ti	$3d^2$	40	Zr	$4d^2$	72	Hf	$5d^2$
23	V	$3d^3$	41	Nb	$4d^3$	73	Ta	$5d^3$
24	Cr	$3d^4$	42	Mo	$4d^4$	74	W	$5d^4$
25	Mn	$3d^5$	43	Tc	$4d^5$	75	Re	$5d^5$
26	Fe	$3d^6$	44	Ru	$4d^6$	76	Os	$5d^6$
27	Co	$3d^7$	45	Rh	$4d^7$	77	Ir	$5d^7$
28	Ni	$3d^8$	46	Pd	$4d^8$	78	Pt	$5d^8$
29	Cu	$3d^9$	47	Ag	$4d^9$	79	Au	$5d^9$
30	Zn	$3d^{10}$	48	Cd	$4d^{10}$	80	Hg	$5d^{10}$
31	Ga	$3d^{10}4s$	49	In	$4d^{10}5s$	80	Tl	$5d^{10}6s$
32	Ge	$3d^{10}4s^2$	50	Sn	$4d^{10}5s^2$	81	Pb	$5d^{10}6s^2$

[a] $(1s^2 2s^2 2p^6 3s^2 3p^6)3d^N = (Ar\ core)3d^N$

[b] $(Ar\ core)(3d^{10}4s^2 4p^6 4d^N) = (Kr\ core)4d^N$

[c] $(Kr\ core)(4d^{10}4f^{14}5s^2 5p^6)5d^N = (Lu^{3+}\ core)5d^N$

TABLE 7.2. ELECTRONIC STRUCTURE OF TRIPLY IONIZED RARE-EARTH IONS

Number	Element	Symbol	Outermost electron shell
57	Lanthanum	La	$4d^{10}4f^0 5s^2 5p^6$
58	Cerium	Ce	$4d^{10}4f^1 5s^2 5p^6$
59	Praseodymium	Pr	$4d^{10}4f^2 5s^2 5p^6$
60	Neodymium	Nd	$4d^{10}4f^3 5s^2 5p^6$
61	Promethium	Pm	$4d^{10}4f^4 5s^2 6p^6$
62	Samarium	Sm	$4d^{10}4f^5 5s^2 5p^6$
63	Europium	Eu	$4d^{10}4f^6 5s^2 5p^6$
64	Gadolinium	Gd	$4d^{10}4f^7 5s^2 5p^6$
65	Terbium	Tb	$4d^{10}4f^8 5s^2 5p^6$
66	Dysprosium	Dy	$4d^{10}4f^9 5s^2 5p^6$
67	Holmium	Ho	$4d^{10}4f^{10} 5s^2 5p^6$
68	Erbium	Er	$4d^{10}4f^{11} 5s^2 5p^6$
69	Thulium	Tm	$4d^{10}4f^{12} 5s^2 5p^6$
70	Ytterbium	Yb	$4d^{10}4f^{13} 5s^2 5p^6$
71	Lutetium	Lu	$4d^{10}4f^{14} 5s^2 5p^6$

(Schiff, 1968), where the $Y_{\ell m}(\hat{r})$ are the spherical harmonics with $\ell = 2$ for the transition-metal ions and $\ell = 3$ for the rare-earth ions. (Remember that $\ell = 0$ for s, $\ell = 1$ for p, and $\ell = 2$ for d electrons.) The radial functions in equation (7.2) are taken to be the same for all the ℓ^N in the ion, while the angular functions, along with the spin of each electron, must form a determinantal function so as to obey the exclusion principle. Depending on whatever determinantal function is chosen, the radial functions can be found by some self-consistent method. These radial functions (Freeman and Watson, 1962; Fraga et al, 1976; Cowan and Griffin, 1976) have been found for the Hund ground states of all the transition-metal and rare-earth ions from cerium (Ce^{3+}) through ytterbium (Yb^{3+}).

The Hund ground state for the transition-metal ions with $N \leq 5$ and the rare-earth ions with $N \leq 7$ is given by assuming that all N spins are parallel and that each angular momentum projection is the maximum allowed by the exclusion principle (in eq (7.2), ℓ is the angular momentum and m is its projection). Thus, the Hund ground state for two electrons is the determinantal function (unnormalized)

$$\alpha(1)Y_{\ell,\ell}(1)\alpha(2)Y_{\ell,\ell-1}(2) - \alpha(1)Y_{\ell,\ell-1}(1)\alpha(2)Y_{\ell,\ell}(2) \quad ,$$

where α is the spin "up" wave function (β = spin down). A convenient notation for such a determinant is

$$\{\overset{+}{\ell}, \ell \overset{+}{-} 1\} \quad , \tag{7.3}$$

where the upper sign is the spin projection (+ = up, - = down) and ℓ and $\ell - 1$ are the z projection of the angular momentum (m in eq (7.2)). Thus, the Hund ground state equation (7.3) for $4f^2$ praseodymium (Pr^{3+}) has total spin, S, and total angular momentum, L, given by

$$S = 1/2 + 1/2 = 1 \quad ,$$

$$L = \ell + \ell - 1 = 2\ell - 1 = 5 \quad (\text{for f}, \ell = 3) \quad .$$

Hence, the ground state is L = 5, with multiplicity of $2S + 1 = 3$. In the so-called Russell-Saunders notation, this state is referred to as ^3H, as given by the following:

Total angular momentum, L, of ion: 0 1 2 3 4 5 6 7 . . .

Russell-Saunders label for the state: S P D F G H I K . . .
(continues alphabetically)

In this notation, the technical reference to such a state is a *term*; other terms for Pr^{3+} are ^1I, ^3P, and ^1S (Condon and Shortley, 1959; Nielson and Koster, 1963). For the ion Ce^{3+}, which has one f electron, the atom notation becomes identical to that of the ion, that is, $\ell = L = 3$ and $s = S = 1/2$ with the single term ^2F. Those ions in the series for $N > 2\ell + 1$ have the same terms as for $N < 2\ell + 1$, and their Hund states can be constructed simply as

$$\{\overset{+}{\ell},\ \ell\overset{+}{-}1,\ \ell\overset{+}{-}2,\ \ldots\ 2\ell\overset{+}{+}1,\ \bar{\ell},\ \ell\bar{-}1,\ \ell\bar{-}2,\ \ldots\ \ell-(\bar{p}+1)\} \qquad (7.4)$$

where the number of electrons is $N = 2\ell + 1 + p$. The ℓ^N shell is completely filled when $N = 2(2\ell + 1)$, which for f electrons is triply ionized lutetium (Lu^{3+}). As an example of equation (7.4), consider triply ionized terbium (Tb^{3+}), which has the $4f^8$ configuration. The determinantal wave function is

$$\{\overset{+}{3}\ \overset{+}{2}\ \overset{+}{1}\ \overset{+}{0}\ \overset{+}{-1}\ \overset{+}{-2}\ \overset{+}{-3}\ \overset{-}{3}\} \quad , \qquad (7.5)$$

where total angular momentum $L = 3$ and total spin angular momentum $S = 6/2 = 3$. Thus, the Hund ground state is ^7F. In all cases, the Hund term has been found to have the lowest energy in atomic systems. In general, the wave functions for the higher terms are very difficult to construct, but sophisticated techniques have been devised for the orderly development of a set of wave functions for each ion having the electronic configuration p^N, d^N, and f^N (Nielson and Koster, 1963). The Hamiltonian given in equation (7.1) has the same value for all terms of the configuration $n\ell^N$; consequently, we ignore H_0 in the future discussion.

7.2 Significant Free-Ion Interactions

A number of interactions within the ion do not depend on the particular solid or are modified when the ion enters a solid. These interactions are termed the free-ion interactions. We discuss several such interactions here.

7.2.1 Coulomb Interaction

The largest contribution to the Hamiltonian for the free ion is the electrostatic interaction of the $n\ell^N$ electrons, which may be written

$$H_1 = \sum_{i>j}^{N} \frac{e^2}{|\vec{r}_{ij}|} \quad , \qquad (7.6)$$

where

$$\vec{r}_{ij} = \vec{r}_i - \vec{r}_j \quad .$$

The matrix elements of this interaction for the state $\{3^+ 2^+\}$ (the ^3H term) of Pr^{3+} are

$$\langle ^3H | H_1 | ^3H \rangle = F^{(0)} - 9F^{(2)} - (17F^{(4)}/363) - (25F^{(6)}/14{,}157) \qquad (7.7)$$

(Judd, 1963), where the $F^{(k)}$ are frequently referred to as the Slater parameters.

The $F^{(k)}$ are radial expectation values given by

$$F^{(k)} = e^2 \int_0^\infty \int_0^\infty \frac{r_<^k}{r_>^{k+1}} \, [R_{n\ell}(r_1) R_{n\ell}(r_2)]^2 r_1^2 \, dr_1 r_2^2 \, dr_2 \quad , \qquad (7.8)$$

where

$$\int_0^\infty R_{n\ell}^2(r) r^2 \, dr = 1 \quad ,$$

$$\frac{r_<}{r_>} = \frac{r_i}{r_j} \qquad \text{if } r_i < r_j \quad \text{and}$$

$$= \frac{r_j}{r_i} \qquad \text{if } r_i > r_j \quad .$$

For the d^N electrons the matrix elements of the Coulomb interaction are given in terms of $F^{(k)}$, while the same interaction for the f^N series is given in terms of new parameters E^k by Nielson and Koster (1963). Nielson and Koster give the matrix elements of the Coulomb interaction in the form

$$\langle \alpha'L'S' | H_1 | \alpha LS \rangle = \delta_{LL'} \delta_{SS'} \sum_k c_k(\alpha',\alpha,L,S) F^{(k)} \quad , \qquad (7.9)$$

and they tabulate the coefficients $c_k(\alpha',\alpha,L,S)$ for each of the states of d^N. Similarly, Nielson and Koster give for the matrix elements of the Coulomb interaction

$$\langle \alpha'L'S' | H_1 | \alpha LS \rangle = \delta_{LL'} \delta_{SS'} \sum_k g_k(\alpha',\alpha,L,S) E^k \quad , \qquad (7.10)$$

and the coefficients $g_k(\alpha',\alpha,L,S)$ are given for each of the states of f^N. In equation (7.9) the values of k are 0, 2, and 4, while in equation (7.10) the k values are 0, 1, 2, and 3. The relation of E^k to $F^{(k)}$ is given in a number of places (for example, Judd, 1963, p 206).

7.2.2 Spin-Orbit Interaction

The second interaction of reasonable magnitude in the free ion is the spin-orbit coupling, which is

$$H_2 = \sum_{i=1}^{N} \zeta(r_i)\vec{\ell}_i \cdot \vec{s}_i \quad , \tag{7.11}$$

where

$$\zeta(r_i) = \frac{\hbar^2}{2m^2c^2}\frac{1}{r_i}\frac{dU(r_i)}{dr_i} \quad .$$

This interaction was derived from relativity theory in the Bohr orbit quantum mechanics, but it is also a natural consequence of a nonrelativistic approximation to the Dirac equation. Values of $F^{(k)}$ and ξ (where $\xi = \langle n\ell|\zeta(r)|n\ell\rangle$) from Hartree-Fock wave functions are given in tables 7.3 to 7.6. In the rare-earth series, the interaction H_2 is quite strong and is in general much larger than the interaction of the rare-earth electrons with the crystal fields.

TABLE 7.3. HARTREE-FOCK VALUES FOR FREE-ION PARAMETERS
FOR DIVALENT IONS WITH $3d^N$ ELECTRONIC
CONFIGURATION (Fraga et al, 1976)

Z	X^{2+}	nd^N	$F^{(2)}$ (cm^{-1})	$F^{(4)}$ (cm^{-1})	ζ_{3d} (cm^{-1})	$\langle r^2\rangle$ (A^2)	$\langle r^4\rangle$ (A^4)
21	Sc	$3d^1$	--	--	85.95	0.8346	1.4997
22	Ti	$3d^2$	67,932	42,357	131.15	0.6716	0.9808
23	V	$3d^3$	74,062	46,171	187.17	0.5677	0.7112
24	C	$3d^4$	79,790	49,726	265.60	0.4910	0.5401
25	Mn	$3d^5$	85,637	53,368	342.85	0.4277	0.4145
26	Fe	$3d^6$	89,877	55,927	441.38	0.3893	0.3527
27	Co	$3d^7$	94,600	58,817	561.21	0.3525	0.2949
28	Ni	$3d^8$	99,392	61,756	703.19	0.3203	0.2478
20	Cu	$3d^9$	--	--	869.65	0.2923	0.2097

TABLE 7.4. HARTREE-FOCK VALUES FOR FREE-ION
PARAMETERS FOR DIVALENT IONS WITH ELECTRONIC
CONFIGURATION $4d^N$ (Fraga et al, 1976)

Z	X^{2+}	nd^N	$F^{(2)}$ (cm^{-1})	$F^{(4)}$ (cm^{-1})	ζ_{4d} (cm^{-1})	$\langle r^2 \rangle$ (A^2)	$\langle r^4 \rangle$ (A^4)
39	Y	$4d^1$	--	--	312.00	1.5737	4.4402
40	Zr	$4d^2$	51,177	33,321	432.03	1.2734	2.8974
41	Nb	$4d^3$	55,682	36,328	566.11	1.0769	2.0761
42	Mo	$4d^4$	59,873	39,117	718.12	0.9316	1.5580
43	Te	$4d^5$	64,052	41,911	891.17	0.8145	1.1907
44	Ru	$4d^6$	67,247	43,978	1081.70	0.7365	0.9869
45	Rh	$4d^7$	70,673	46,224	1299.11	0.6656	0.8126
46	Pd	$4d^8$	74,108	48,480	1544.04	0.6045	0.6744
47	Ag	$4d^9$	--	--	1820.08	0.5516	0.5644

TABLE 7.5. HARTREE-FOCK VALUES FOR FREE-ION
PARAMETERS FOR DIVALENT IONS WITH ELECTRONIC
CONFIGURATION $5d^N$ (Fraga et al, 1976)

Z	X^{2+}	nd^N	$F^{(2)}$ (cm^{-1})	$F^{(4)}$ (cm^{-1})	ζ_{5d} (cm^{-1})	$\langle r^2 \rangle$ (A^2)	$\langle r^4 \rangle$ (A^4)
71	Lu	$5d^1$	--	--	1390.74	1.6197	4.6324
72	Hf	$5d^2$	50,350	33,000	1773.59	1.3646	3.2437
73	Ta	$5d^3$	54,008	35,526	2170.38	1.1926	2.4612
74	W	$5d^4$	57,369	37,840	2594.40	1.0610	1.9385
75	Re	$5d^5$	60,702	40,149	3052.92	0.9510	1.5467
76	Os	$5d^6$	63,123	41,766	3530.75	0.8779	1.3277
77	Ir	$5d^7$	65,755	43,550	4056.26	0.8087	1.1289
78	Pt	$5d^8$	68,388	45,344	4626.25	0.7474	0.9649
79	Au	$5d^9$	--	--	5247.58	0.6930	0.6646

TABLE 7.6. NONRELATIVISTIC HARTREE-FOCK INTEGRALS FOR TRIPLY IONIZED
RARE-EARTH IONS (Fraga et al, 1976)

N	R^{3+}	$F^{(2)}$ (cm^{-1})	$F^{(4)}$ (cm^{-1})	$F^{(6)}$ (cm^{-1})	ζ (cm^{-1})	$\langle r^2 \rangle$ (A^2)	$\langle r^4 \rangle$ (A^4)	$\langle r^6 \rangle$ (A^6)
1	Ce	--	--	--	778.14	1.1722	3.0818	15.549
2	Pr	105,120	66,213	47,718	919.16	1.0632	2.5217	11.492
3	Nd	109,731	69,165	49,860	1069.87	0.97822	2.1317	8.9525
4	Pm	113,640	71,641	51,647	1228.24	0.91401	1.8701	7.4224
5	Sm	117,222	73,893	53,269	1397.79	0.86059	1.6700	6.3365
6	Eu	120,885	76,204	54,937	1583.54	0.81064	1.4989	5.3886
7	Gd	124,644	78,585	56,655	1786.68	0.76368	1.3267	4.5589
8	Tb	127,137	80,091	57,722	1990.51	0.73523	1.2508	4.2673
9	Dy	129,960	81,829	58,962	2214.87	0.70484	1.1644	3.902
10	Ho	132,929	83,670	60,281	2458.58	0.67481	1.0788	3.5296
11	Er	135,859	85,486	61,580	2719.76	0.64714	1.0028	3.2111
12	Tm	138,754	87,276	62,864	2999.22	0.62154	0.93514	2.9355
13	Yb	--	--	--	3299.82	0.59678	0.87030	2.6713

Consequently, it is convenient to perform all the calculations in a set of basis functions in which H_1 and H_2 are diagonal. The set of functions that achieves this is the total angular momentum function $|JM_J\rangle$, where the total angular momentum operator $\vec{J} = \vec{L} + \vec{S}$.

The spin-orbit interaction H_2 given in equation (7.11) commutes with the total angular momentum and, consequently, since H_1 also commutes with $\vec{J}^2$, the wave functions can be characterized by the eigenvalues of $\vec{J}^2$ and J_z. That is, we can write ψ_{JM} or $|JM\rangle$ for the wave functions with

$$\vec{J}^2 |JM\rangle = J(J+1)|JM\rangle \quad \text{and}$$

$$(7.12)$$

$$J_z |JM\rangle = M|JM\rangle \quad .$$

For any term of given L and S (eigenvalues of $\vec{L}^2$ and $\vec{S}^2$), the values of J are restricted to

$$|L - S| \leq J \leq |L + S| \quad .$$

Then the wave functions are customarily written ψ_{JMLS} or $|JMLS>$, and we have

$$\vec{L}^2|JMLS> = L(L+1)|JMLS> \ ,$$

$$\vec{S}^2|JMLS> = S(S+1)|JMLS> \ ,$$

$$\vec{J}^2|JMLS> = J(J+1)|JMLS> \ , \qquad (7.13)$$

$$J_z|JMLS> = M|JMLS> \ ;$$

and

$$<J'M'L'S'|H_1+H_2|JMLS> \propto \delta_{JJ'}\delta_{MM'} \ . \qquad (7.14)$$

As implied in equation (7.14), the energy H_1 and H_2 is independent of M, or each J level of the free ion is 2J+1-fold degenerate. The matrix elements in equation (7.14) do not vanish generally for L' = L ± 1 and S' = S ± 1; thus, L and S are not strictly good quantum numbers. Nevertheless, the energy levels are labeled as though they were, as in the Russell-Saunders notation, $^{2S+1}L_J$. An example of the energy levels for the $4f^2$ configuration of the free ion (Pr^{3+}) is given in table 7.7. Also included is the same ion, Pr^{3+}, in the host materials lanthanum trichloride ($LaCl_3$) and lanthanum trifluoride (LaF_3).

The results in table 7.7 are interesting in that they show that most of the energy levels observed in the free ion are lowered when the ion is embedded in a solid. This shift in the energy levels is a general effect and is not restricted to Pr^{3+}, but exists in all the rare-earth ions where a comparison with energy levels of the free ion can be made. In fact, this shift has been observed by Low (1958a,b) in ions with an unfilled d shell. The first explanation of this shift in energies was by Morrison et al (1967), who showed that if the ion under investigation was assumed to be embedded in a solid of homogeneous dielectric constant, ε, then a decrease in the Slater integrals is given by

$$\Delta F^{(k)} = -e^2(\varepsilon-1)(k+1)<r^k>^2/\{b^{2k+1}[\varepsilon+k(\varepsilon+1)]\} \ ,$$

$$(7.15)$$

where b is the radius of a fictitious cavity surrounding the rare-earth ion. The result

TABLE 7.7. FREE-ION ENERGY LEVELS OF TRIPLY IONIZED PRASEODYMIUM AND CORRESPONDING CENTROIDS IN TWO CRYSTALS (all in cm^{-1}) (Dieke, 1968)

[LS]J	Free	$LaCl_3$	LaF_3
$^3H_4$	0	0	0
$^3H_5$	2,152	2,119	2,163
$^3H_6$	4,389	4,307	4,287
$^3F_2$	4,997	4,848	5,015
$^3F_3$	6,415	6,248	6,368
$^3F_4$	6,855	6,684	6,831
$^1G_4$	9,921	9,704	9,801
$^1D_2$	17,334	16,640	16,847
$^3P_0$	21,390	20,385	20,727
$^3P_1$	22,007	21,987	21,314
$^1I_6$	22,212	21,327	--
$^3P_2$	23,160	22,142	22,546
$^1S_0$	50,090	48,710	46,786

given in equation (7.15) was first successfully applied to Co^{++} in $MgAl_2O_4$. Later, Newman (1973) showed that the shift in $F^{(k)}$ given in equation (7.15) was sufficiently large to predict the shifts in the energy levels for rare-earth ions. More recently, Morrison (1980) derived the result

$$\Delta F^{(k)} = - \sum_i \frac{\alpha_i Z_i e^2}{R_i^{2k+4}} (k+1)\langle r^k \rangle^2 \quad , \tag{7.16}$$

where α_i is the polarizability of the Z_i ligands at R_i and $\langle r^k \rangle$ is the radial expectation value of r^k. The result given in equation (7.16) is believed to be more fundamental than that of equation (7.15) because the latter explicitly accounts for the local coordination of the rare-earth metal ion. Morrison (1980) gives a predicted shift in the spin-orbit parameter, ζ, but because of the smallness of the predicted shift and the errors in the fitting of the experimental data, no comparisons were made.

Because of the lack of experimental data on the free-ion spectra of rare-earth ions, measurement of the shift in the Slater integrals is possible only for Pr^{3+}. The experimental $F^{(k)}$ for triply ionized rare earths in $LaCl_3$ have been obtained by Carnall et al (1978), and these results are given in table 7.8, and the corresponding experimental values for the transition-metal ions are given in table 7.9. These data can be used in conjunction with equation (7.16) to obtain results that can perhaps be applied to an arbitrary host material to predict a priori the energy level shift of that host.

TABLE 7.8.   FREE-ION PARAMETERS FOR TRIPLY IONIZED
RARE-EARTH IONS IN $LaCl_3$ OBTAINED FROM FITTING
EXPERIMENTAL DATA (all in cm^{-1}) (Carnall et al, 1978)
Values in parentheses were not varied in the fitting

Ion	F^2	F^4	F^6	α	β	γ	ζ
Pr	68,368	50,008	32,743	22.9	-674	(1520)	744
Nd	71,866	52,132	35,473	22.1	-650	1586	880
Pm	75,808	54,348	38,824	21.0	-645	1425	1022
Sm	78,125	56,809	40,091	21.6	-724	(1700)	1168
Eu	84,399	60,343	41,600	16.8	(-640)	(1750)	1331
Gd	85,200	60,399	44,847	(19)	(-643)	1644	(1513)
Tb	90,012	64,327	42,951	17.5	(-630)	(1880)	1707
Dy	92,750	65,699	45,549	17.2	-622	1881	1920
Ho	95,466	67,238	46,724	17.2	-621	2092	2137
Er	98,203	69,647	49,087	15.9	-632	(2017)	2370

TABLE 7.9. EXPERIMENTAL VALUES OF FREE-ION
PARAMETERS (cm^{-1}) FOR DIVALENT $3d^N$ ELECTRONIC
CONFIGURATION (Uylings et al, 1984)

Z	X^{2+}	nd^N	F$^{(2)}$	F$^{(4)}$	ζ_d	α	β
22	Ti	$3d^2$	54,927	32,206	118	20.52	-466.2
23	V	$3d^3$	59,924	36,268	170	22.90	-480.8
24	Cr	$3d^4$	64,798	40,288	231	25.83	-509.2
25	Mn	$3d^5$	69,485	44,305	316	29.20	-537.1
26	Fe	$3d^6$	74,282	48,241	422	33.21	-533.2
27	Co	$3d^7$	78,906	52,227	536	37.48	-532.48
28	Ni	$3d^8$	83.514	56,164	668	42.49	-554.9

7.2.3 Interconfigurational Interaction

An interaction that has been frequently used in fitting the "free" ion levels of a rare-earth ion or a transition-metal ion in a crystal is the so-called interconfigurational mixing or the Trees interaction. For the rare-earth ions this interaction has been parametrized by Wybourne and Rajnak (Wybourne, 1965) and is

$$H_{10} = \alpha L(L+1) + \beta G(G_2) + \gamma G(R_7) \quad , \qquad (7.17)$$

where α, β, and γ are parameters adjusted to fit the experimental data. The operator $G(G_2)$ is the Casimir operator for the group G_2, and $G(R_7)$ is the similar operator for R_7 (note that $\vec{L}^2 = L(L+1)$ is the Casimir's operator for the group R_3). The values for these operators for all the states are tabulated by Wybourne (1965, p 73). The values for the state of f^2 are given in table 7.10. The values of α, β, and γ obtained by fitting experimental data for the rare-earth ions are given in table 7.8. To my knowledge, no successful attempts to derive theoretical values of α, β, and γ have been published. For the transition-metal ions the Trees interaction in equation (7.17) uses α, but β either multiplies $G(R_5)$ or the seniority operator Q (Wybourne, 1965).

TABLE 7.10. EIGEN-
VALUES OF CASIMIR'S
OPERATORS FOR STATE
OF f^2

State	α	13β	5γ
^3P	2	12	5
^3F	12	6	5
^3H	30	12	5
^1S	0	0	0
^1D	6	14	7
^1G	20	14	7
^1I	42	14	7

7.2.4. Other Interactions

Many other interactions are considered in the free ion, such as spin-other-orbit, orbit-orbit, and configuration interaction. All these, to a greater or lesser extent, improve the fit of theoretical energy levels to the experimental data. We omit these interactions from further discussion since

H_1, H_2, and H_{10} give a sufficient representation of the free ion for our purposes here. However, we list a number of interactions including the above which have been considered by various research workers (Wortman et al, 1973):

H_1 = the Coulomb interaction

H_2 = the spin-orbit interaction

H_3 = the crystal-field interaction

H_4 = the interaction with a magnetic field (Zeeman effect)

H_5 = the hyperfine interaction

H_6 = the spin-spin interaction

H_7 = the nuclear quadrupole interaction

H_8 = the spin-other-orbit interaction

H_9 = the orbit-orbit interaction

H_{10} = the interconfigurational interaction

H_{11} = the spin-crystal-field interaction

The notation listed above is that of Judd (1963), with a few obvious additions.

7.3 Summary

We have considered the Coulomb interaction, H_1, and the spin-orbit interaction, H_2, for the configuration $n\ell^N$ in the free ion. The wave functions that are chosen as a basis for diagonalization of these interactions are $|JMLS\rangle$, and the resulting energy levels are labeled according to the Russell-Saunders notation as given in section 7.1. This same notation (plus additional quantum numbers) is used for describing an ion in a crystal. The values of $\langle r^k \rangle$ that are needed in equations (7.15) and (7.16) are given in table 7.6 for the rare-earth ions, and tables 7.2, 7.3, and 7.4 give the corresponding values for the doubly ionized nd^N ions. The wave functions used for the calculation of the energy levels of rare-earth and transition-metal ions in a solid will be the combination that simultaneously diagonalizes H_1 and H_2. While this process is not a good one for the $3d^N$ configuration, it is better for the $4d^N$ and $5d^N$ configurations and is excellent for the triply ionized rare-earth ions.

7.4 Problems

1. We have the tensor operator T_{kq} given by

$$T_{kq} = \sum_{i=1}^{N} C_{kq}(i) \ .$$

77

(a) For $N = 2$ evaluate the matrix elements of T_{kq} by using the states given by equation (7.3). That is, show that

$$\{\hat{\ell},\ell\overset{+}{-}1\}^{*}T_{kq}\{\hat{\ell},\ell\overset{+}{-}1\} = \left[\langle\ell\ell|C_{kq}|\ell\ell\rangle + \langle\ell\ell-1|C_{kq}|\ell\ell-1\rangle\right]\delta_{q0} \quad .$$

(b) By application of the Wigner-Eckart theorem to the problem in (a) we have

$$\{\hat{\ell},\ell\overset{+}{-}1\}^{*}T_{kq}\{\hat{\ell},\ell\overset{+}{-}1\} = \langle L(L)k(0)|L(L)\rangle\ \langle LS\|T_{k}\|LS\rangle$$

with $L = 2\ell - 1$.

Using this result and the result in (a) show that

$$\langle LS\|T_{k}\|LS\rangle = \frac{\langle\ell(0)k(0)|\ell(0)\rangle}{\langle L(L)k(0)|L(L)\rangle}\left[\langle\ell(\ell)k(0)|\ell(\ell)\rangle + \langle\ell(\ell-1)k(0)|\ell(\ell-1)\rangle\right]$$

where

$$\langle L(L)k(0)|L(L)\rangle = \left[\frac{(2L+1)(2L)!(2L)!}{(2L+k+1)(2L+k)!(2L-k)!}\right]^{1/2}$$

and

$$\langle\ell(\ell)k(0)|\ell(\ell)\rangle = \text{as above with } L = \ell.$$

Show also that

$$\langle\ell(\ell-1)k(0)|\ell(\ell-1)\rangle = \left[\frac{(2\ell+1)(2\ell-1)!(2\ell-1)!}{(2\ell-k)!}\right]^{1/2}\left[2\ell-k-k^{2}\right] \quad .$$

2. By extending the results obtained in problem 1 to the operator

$$T_{kq} = \sum_{i=1}^{N} C_{kq}(i) \quad ,$$

show that

$$\langle LLSS|T_{k0}|LLSS\rangle = \sum_{p=0}^{N} \langle\ell(\ell-p)k(0)|\ell(\ell-p)\rangle\ \langle\ell(0)k(0)|\ell(0)\rangle \quad ,$$

where $L = N\ell - \dfrac{N(N-1)}{2}$, $S = \dfrac{N}{2}$, $N \leq 2\ell + 1$.

3. For $N = 2\ell + 1$ what is the value of the following sum?

$$S_N(k) = \sum_{p=0}^{2\ell} \langle \ell(\ell-p)k(0)|\ell(\ell-p)\rangle$$

4. The recursion relation

$$[k(k+1) - 2\ell(\ell+1) + 2m^2] \langle \ell(m)k(0)|\ell(m)\rangle$$

$$= -(\ell+m)(\ell-m+1) \langle \ell(m-1)k(0)|\ell(m-1)\rangle - (\ell-m)(\ell+m+1) \langle \ell(m+1)k(0)|\ell(m+1)\rangle$$

can be used to reduce the number of Clebsch-Gordon coefficients in problems 1 and 2. Obtain this recursion formula (see Rose, 1957, chapter III).

5. For $\langle \ell(0)k(0)|\ell(0)\rangle$ derive the recursion formula (eq (2.10))

$$\langle \ell|C_{k+2}|\ell\rangle = - \frac{k+1}{k+2} \left[\frac{(2\ell+k+2)(2\ell-k)}{(2\ell-k-1)(2\ell+k+3)} \right]^{1/2} \langle \ell|C_k|\ell\rangle$$

for even k. Thus since

$$\langle \ell|C_0|\ell\rangle = 1 \quad ,$$

any given $k\langle \ell|C_k|\ell\rangle$ can be evaluated algebraicly.

7.5 Annotated Bibliography and References

Bishton, S. S., and D. J. Newman (1970), Parametrization of the Correlation Crystal Field, J. Phys. C 3, 1753.

Carnall, W. T., H. Crosswhite, and H. M. Crosswhite (1978), Energy Level Structure and Transition Probabilities in the Spectra of the Trivalent Lanthanides in LaF_3, Argonne National Laboratory, ANL-78-XX-95.

Condon, E. U., and H. Odabasi (1980), Atomic Structure, Cambridge University Press, Cambridge, U.K., chapters 5 and 8.

Condon, E. U., and G. H. Shortley (1959), The Theory of Atomic Spectra, Cambridge University Press, Cambridge, England.

Cowan, R. D., and D. C. Griffin (1976), Approximate Relativistic Corrections to Atomic Radial Wave Functions, J. Opt. Soc. Am. 66, 1010.

Dieke, G. H. (1968), Spectra and Energy Levels of Rare Earth Ions in Crystals, Interscience, New York, NY, p 200.

Dieke, G. H., and H. M. Crosswhite (1963), The Spectra of the Doubly and Triply Ionized Rare Earths, Appl. Opt. $\underline{2}$, 675.

Fraga, S., K. M. S. Saxena, and J. Karwowski (1976), Physical Science Data 5, Handbook of Atomic Data, Elsevier, New York, NY.

Freeman, A. J., and R. E. Watson (1962), Theoretical Investigation of Some Magnetic and Spectroscopic Properties of Rare-Earth Ions, Phys. Rev. $\underline{127}$, 2058.

Judd, B. R. (1963), Operator Techniques in Atomic Spectroscopy, McGraw-Hill, New York, NY. Chapters 1, 2, 3, and 4 are very pertinent to this section.

Low, W. (1958a), Paramagnetic and Optical Spectra of Divalent Nickel in Cubic Crystalline Fields, Phys. Rev. $\underline{109}$, 247.

Low, W. (1958b), Paramagnetic and Optical Spectra of Divalent Cobalt in Cubic Crystalline Fields, Phys. Rev. $\underline{109}$, 256.

Morrison, C. A. (1980, January 15), Host Dependence of the Rare-Earth Ion Energy Separation $4F^N - 4F^{N-1}n\ell$, J. Chem. Phys. $\underline{72}$, 1001.

Morrison, C. A., and R. P. Leavitt (1982), Spectroscopic Properties of Triply Ionized Lanthanides in Transparent Host Materials, in Volume 5, Handbook of the Physics and Chemistry of Rare Earths, ed. by K. A. Gschneidner, Jr., and L. Eyring, North-Holland Publishers, New York, NY.

Morrison, C. A., D. R. Mason, and C. Kikuchi (1967), Modified Slater Integrals for an Ion in a Solid, Phys. Lett. $\underline{24A}$, 607.

Newman, D. J. (1973), Slater Parameter Shifts in Substituted Lanthanide Ions, J. Phys. Chem. Solids $\underline{34}$, 541.

Nielson, C. W., and G. F. Koster (1963), Spectroscopic Coefficients for p^n, d^n and f^n Configurations, MIT Press, Cambridge, MA.

Schiff, L. I. (1968), Quantum Mechanics, 3rd ed., McGraw-Hill, New York, NY.

Trees, R. E. (1964), $4f^3$ and $4f^25d$ Configuration of Doubly Ionized Praseodymium (Pr III), J. Opt. Soc. Am. $\underline{54}$, 651.

Uylings, P. H. M., A. J. J. Raassen, and J. F. Wyart (1984), Energies of N Equivalent Electrons Expressed in Terms of Two-Electron Energies and Independent Three-Electron Parameters: A New Complete Set of Orthogonal Operators: II.--Application of $3d^N$ Configurations, J. Phys. $\underline{B17}$, 4103.

Wortman, D. E., R. P. Leavitt, and C. A. Morrison (1973, December), Analysis of the Ground Configuration of Trivalent Thulium in Single-Crystal Yttrium Vanadate, Harry Diamond Laboratories, HDL-TR-1653.

Wybourne, B. G. (1965), Spectroscopic Properties of Rare Earths, Wiley, New York, NY, p 72.

8. CRYSTAL-FIELD INTERACTIONS--PHENOMENOLOGICAL THEORY OF CRYSTAL FIELDS

8.1 Discussion

In the presence of a crystal field we take the interaction of an ion whose electronic configuration is $n\ell^N$ as

$$H_3 = \sum_{kq} B_{kq}^* \sum_{i=1}^{N} C_{kq}(\hat{r}_i) \quad , \tag{8.1}$$

where the C_{kq} are unnormalized spherical harmonics given by

$$C_{kq}(\hat{r}) = [4\pi/(2k+1)]^{1/2} Y_{kq}(\hat{r}) \quad .$$

The use of the C_{kq} in expressions for electronic interactions (along with other shorthand notation that we will not use) is practically universal. The number of terms in equation (8.1) that need to be considered is limited by the symmetry of the site occupied by the particular ion. Also, since we will be discussing only the $n\ell^N$ configuration, k is limited to values of 4 for $\ell = 2$ and 6 for $\ell = 3$. This limitation arises because, independent of the basis chosen, individual matrix elements of C_{kq} will have to be considered, and these are such that $\langle\ell|C_k|\ell\rangle = 0$ if $k > 2\ell$.

For our purposes in the discussion of the nonvanishing B_{kq} of equation (8.1), it is sufficient to consider a single electron; thus, the Hamiltonian we consider is

$$H_3 = \sum_{kq} B_{kq}^* C_{kq} \quad . \tag{8.2}$$

Since H_3 must be hermitian, we have the property that $B_{k-q} = (-1)^q B_{kq}^*$, which is the same as the C_{kq} given in equation (1.25). The basic assumption of group theory is that H_3 is invariant under all the operations of the groups under consideration. That is, we shall assume that

$$O_\lambda H_3 = H_3 \tag{8.3}$$

where O_λ is any operation of the group. Using equations (8.3) and (8.2) we have

$$O_\lambda H_3 = \sum_{kq} B^*_{kq} O_\lambda C_{kq} \quad , \tag{8.4}$$

where we assume that O_λ operates only on the electron coordinates. The B_{kq} contain the dependence on r, but since r is invariant for all the 32 point groups, there is no loss in generality. Thus, we need only consider the results of $O_\lambda C_{kq}$ for all O_λ in a particular group. The lowest group C_1 contains the identity operator only; consequently all B_{kq} are allowed. The group C_i contains the inversion operation I, and $IC_{kq} = (-1)^k C_{kq}$, so that equations (8.3) and (8.4) give $B_{kq} = 0$ for all odd k, while all B_{kq} with even k are allowed. The group C_2 contains the operation C_2, a rotation about the z-axis by 180°; thus $C_2 C_{kq} = (-1)^q C_{kq}$. Consequently for the group C_2 all k values are allowed and only the B_{kq} with even q exist.

The group C_s has the symmetry operator σ_h, given in problem 2 of section 1. That is,

$$\sigma_h C_{kq} = (-1)^{k+q} C_{kq} \quad ,$$

which, for even k, gives the same nonvanishing B_{kq} as C_2, but for odd k, q must be odd, giving B_{11}, B_{31}, and B_{33} for odd k < 5.

We next consider the two groups C_4 and S_4. The symmetry operators for the C_4 group are C_4 (a rotation about the z-axis by 90°), C_2, and C_4^3 or C_4^{-1} ($C_4^4 = E$).

For the first of these operations and using table 1.1 we have

$$C_4 C_{kq} = e^{i\pi q/2} C_{kq} \quad , \tag{8.5}$$

and no new restrictions are imposed by the symmetry operations C_2 and C_4^3. The nonvanishing crystal-field parameters are

$$B_{kq} = 0$$

($|q| < k$ and k is any positive integer), unless q = 0, ±4, ±8, ±12, ...

For the S_4 group, the symmetry operator S_4 can be written $S_4 = IC_4^{-1}$, so that

$$S_4 C_{kq} = (-1)^k e^{-i\pi q/2} C_{kq} \quad . \tag{8.6}$$

Thus, for even k, S_4 has the same nonvanishing B_{kq} as C_4, but for odd k,

$$B_{kq} = 0$$

unless q = ±2, ±6 ... with odd k and $|q| \leq k$. This condition gives rise to the nonvanishing B_{kq} as B_{32}, B_{52}, B_{72}, B_{76} ...

As a final example we consider the group D_2. The symmetry operations for this group are $C_2(x)$, $C_2(y)$, and $C_2(z)$. From problem 2 of section 1 we have

$$C_2(x)C_{kq} = (-1)^k C_{k-q} \quad,$$

$$C_2(y)C_{kq} = (-1)^{k+q} C_{k-q} \quad, \tag{8.7}$$

$$C_2(z)C_{kq} = (-1)^q C_{kq} \quad.$$

The last relation in equation (8.7) requires that q be even; from this result the first two relations are identical. If the first relation in equation (8.7) is substituted into equation (8.4) and the summing index q is changed to $-q$, we have

$$B_{kq} = (-1)^k B_{k-q} \quad. \tag{8.8}$$

Since $B_{k-q} = (-1)^q B_{kq}^*$ (see the discussion of eq (8.2)), and since q is even,

$$B_{kq} = (-1)^k B_{kq}^* \quad. \tag{8.9}$$

The result given in equation (8.9) requires that B_{kq} be real for even k and imaginary for odd k; also $B_{k0} = 0$ for all odd k, since $B_{k0} = B_{k0}^*$ in general.

Using the above techniques and table 1.1 we obtain the results given in tables 8.1 and 8.2, which are sufficient for d electrons; for f electrons, however, the tables must be extended to $k = 6$. For the cubic groups the operations are more involved; the crystal-field Hamiltonian is given in table 8.3 for $k \leq 6$, which is sufficient for the rare earths.

Sometimes in the O group the (111) cubic axis is chosen as the z-axis; the crystal field is then given as

$$H_{CEF}^{[111]} = B_{40}[C_{40} \mp \sqrt{10/7} \, (C_{43} - C_{4-3})]$$

$$\tag{8.10}$$

$$+ B_{60}[C_{60} \pm \sqrt{35/96} \, (C_{63} - C_{6-3}) + \sqrt{77/192} \, (C_{66} + C_{6-6})] \quad.$$

It may be of interest to note that the lowest odd-k term in the crystal field for the O group is (from Polo, 1961)

$$B_{94}[(C_{94} - C_{9-4}) - \sqrt{7/17} \, (C_{98} - C_{9-8})] \quad.$$

83

TABLE 8.1. ALLOWED VALUES OF B_{kq} FOR POINT GROUPS 3 THROUGH 15[a]

No.[b]	Group[c]	B_{10}	B_{11} Re	Im	B_{20}	B_{22} Re	Im	B_{30}	B_{31} Re	Im	B_{32} Re	Im	B_{33} Re	Im	B_{40}	B_{42} Re	Im	B_{44} Re	Im
3	C_2	X	0	0	X	X	X	X	0	0	X	X	0	0	X	X	X	X	X
4	C_s	0	X	X	X	X	X	0	X	X	0	0	X	X	X	X	X	X	X
5	C_{2h}	0	0	0	X	X	X	0	0	0	0	0	0	0	X	X	X	X	X
6	D_2	0	0	0	X	X	0	0	0	0	0	X	0	0	X	X	0	X	0
7	C_{2v}	X	0	0	X	X	0	X	0	0	X	0	0	0	X	X	0	X	0
8	D_{2h}	0	0	0	X	X	0	0	0	0	0	0	0	0	X	X	0	X	0
9	C_4	X	0	0	X	0	0	X	0	0	0	0	0	0	X	0	0	X	X
10	S_4	0	0	0	X	0	0	0	0	0	X	X	0	0	X	0	0	X	X
11	C_{4h}	0	0	0	X	0	0	0	0	0	0	0	0	0	X	0	0	X	X
12	D_4	0	0	0	X	0	0	0	0	0	0	0	0	0	X	0	0	X	0
13	C_{4v}	X	0	0	X	0	0	X	0	0	0	0	0	0	X	0	0	X	0
14	D_{3d}	0	0	0	X	0	0	0	0	0	X	0	0	0	X	0	0	X	0
15	D_{4h}	0	0	0	X	0	0	0	0	0	0	0	0	0	X	0	0	X	0

[a] An X indicates the presence of B_{kq} and a 0 its absence. Missing B_{kq} are 0.
[b] The number is that given in Koster et al (1963).
[c] Schoenflies notation. For the relation to other notations, see Koster et al (1963).

TABLE 8.2. ALLOWED VALUES OF B_{kq} FOR
POINT GROUPS 16 THROUGH 27[a]

No.[b]	Group[c]	B_{10}	B_{20}	B_{30}	B_{33} Re	Im	B_{40}	B_{43} Re	Im
16	C_3	X	X	X	X	X	X	X	X
17	C_{3i}	0	X	0	0	0	X	X	X
18	D_3	0	X	0	0	X	X	X	0
19	C_{3v}	X	X	X	X	0	X	X	0
20	D_{3d}	0	X	0	0	0	X	X	0
21	C_6	X	X	X	0	0	X	0	0
22	C_{3h}	0	X	X	X	0	X	0	0
23	C_{6h}	0	X	0	0	0	X	0	0
24	D_6	0	X	0	0	0	X	0	0
25	C_{6v}	X	X	X	0	0	X	0	0
26	D_{3h}	0	X	X	0	0	X	0	0
27	D_{6h}	0	X	0	0	0	X	0	0

[a] An X indicates the presence of B_{kq} and a 0 its absence. Missing B_{kq} are 0.
[b] The number is that given in Koster et al (1963).
[c] Schoenflies notation. For the relation to other notations, see Koster et al (1963).

TABLE 8.3   ALLOWED VALUES OF B_{kq} $(k \leq 5)$
FOR CUBIC GROUPS 28 THROUGH $32^{a,b}$

No.	Group	H_{CEF}
28	T	$B_{32}(C_{32} + C_{3-2}) + B_{40}[C_{40} \pm \sqrt{5/14}\,(C_{44} + C_{4-4})]$
		$+ B_{60}[C_{60} \mp \sqrt{7/2}\,(C_{64} + C_{6-4})]$
29	T_h	Same as 28 but $B_{32} = 0$
30	O^b	Same as 29
31	T_d	Same as 28
32	O_h	Same as 30

[a]*In all cases the signs of the B_{4q} and B_{6q} parameters are correlated.*
[b]*The z-axis is chosen along the (001) cubic axis (see discussion of eq (8.10)).*

A method of checking the results given in tables 8.1, 8.2, and 8.3 is by the use of the full rotation compatibility tables of Koster et al (1963). These tables can be used to determine the number of times the identity representation appears in a given C_{kq}. Actually the method can be used to find the C_{kq} that form basis functions for the identity representation and can be extended to other representations of the single group. For a given k in C_{kq}, we look for the number of Γ_1's corresponding to $D_k^{\pm}$, and for the C_{kq} we use the plus sign for even k and the minus sign for odd k $(IC_{kq} = (-1)^k C_{kq})$.

As an example we consider the C_2 group. Using table 13 of Koster et al (1963), we have $D_1^{\pm} = \Gamma_1 + 2\Gamma_2$, or one Γ_1 which, from our previous work, corresponds to C_{10}. Thus we have the parameter B_{10}. For $D_2^{\pm}$ there are three Γ_1's, which correspond to B_{20} and B_{22}; since B_{22} is complex there are three independent constants.

As a second example we choose the group C_{2v}. From table 22 of Koster et al, for D_1^- we have one Γ_1 which corresponds to B_{10}; for D_2^+ we have two Γ_1's, which correspond to B_{20} and ReB_{22}. Normally there would be three B_{2q}, but from our previous results we know that B_{22} is real. This process can be repeated for all the point groups; the results are summarized in table 8.4.

The results presented in tables 8.1 through 8.3 were given assuming a single electron; however, it is generally assumed that the radial dependence which is contained in B_{kq} is the same for all N electrons in a single configuration $n\ell^N$. This assumption is inherent in the crystal-field interaction given in equation (8.1).

Despite all the restrictions imposed by the symmetry operations used above, there exists still one more restriction (possibly more in some groups) that can be imposed on the B_{kq}. This would be apparent if we were to consider a particular model for the computation of the B_{kq}. Any such model would be based on a coordinate system embedded in the crystal and could be used to determine, say, B_{kq} for a particular k and q. Now if we wished to change to a second coordinate system obtained from the first by a simple rotation about the z-axis, we would obtain, say, B'_{kq}, and the two sets of parameters are related by

$$B'_{kq} = e^{iq\phi} B_{kq} \qquad (8.11)$$

where ϕ is the angle of rotation.

This result, equation (8.11), shows that we can choose ϕ such that we can make one of the B_{kq} real and positive. To show this we assume that B'_{KQ} is the parameter we would like to be real and positive.

First assume

$$B_{KQ} = \alpha - i\beta \quad . \qquad (8.12)$$

Then, from equation (4.10), we obtain

$$B'_{KQ} = (\alpha \cos Q\phi + \beta \sin Q\phi) + i(\alpha \sin Q\phi - \beta \cos Q\phi) \quad .$$

TABLE 8.4. NUMBER OF TIMES IDENTITY REPRESENTATION APPEARS IN C_{kq}

No.	Group	\multicolumn{6}{}{The number of Γ_1's for}
		k = 1	2	3	4	5	6
3	C_2	1	3	3	5	5	7
4	C_s	2	3	4	5	6	7
5	C_{2h}	0	3	0	5	0	7
6	D_2	0	2	1	3	2	4
7	C_{2v}	1	2	2	3	3	4
8	D_{2h}	0	2	0	3	0	4
9	C_4	1	1	1	3	3	3
10	S_4	0	1	2	3	2	3
11	C_{4h}	0	1	0	3	0	3
12	D_4	0	1	0	2	1	2
13	C_{4v}	1	1	1	2	2	2
14	D_{2d}	0	1	1	2	1	2
15	D_{4h}	0	1	0	2	0	2
16	C_3	1	1	3	3	3	5
17	C_{3i}	0	1	0	3	0	5
18	D_3	0	1	1	2	1	3
19	C_{3v}	1	1	2	2	2	3
20	D_{3d}	0	1	0	2	0	3
21	C_6	1	1	1	1	1	3
22	C_{3h}	0	1	2	1	2	3
23	C_{6h}	0	0	0	0	0	3
24	D_6	0	1	0	1	0	2
25	C_{6v}	1	1	1	1	1	2
26	D_{3h}	0	1	1	1	1	2
27	D_{6h}	0	1	0	1	0	2
28	T	0	0	1	1	0	2
29	T_h	0	0	0	1	0	2
30	O	0	0	0	1	0	1
31	T_d	0	0	1	1	0	1
32	O_h	0	0	0	1	0	1

The imaginary part of B'_{KQ} vanishes if

$$\tan Q\phi_0 = \frac{\beta}{\alpha} \quad \text{or} \tag{8.13}$$

$$\phi_0 = \frac{1}{Q} \tan^{-1}\left(\frac{\beta}{\alpha}\right) + \frac{p\pi}{Q} \tag{8.14}$$

where $p = 0, \pm 1$.

The real part of B'_{KQ} can be made positive for an appropriate choice of p. The complete set of B'_{kq} is then obtained from equation (8.11) with the appropriate ϕ_0 to give

$$B'_{kq} = e^{iq\phi_0} B_{kq} \quad . \tag{8.15}$$

In most of our work, the lowest B_{kq} for which k is even and $q \neq 0$ has been chosen as real and positive. Thus, for C_2 point symmetry, the twofold crystal-field interaction is written

$$B_{20}C_{20} + B_{22}\left(C_{22} + C_{2-2}\right) \tag{8.16}$$

with B_{22} real and positive.

This reduction of the number of phenomenological crystal-field parameters needed to fit the experimental data does not help much in the low-symmetry point groups (point groups 1 through 8) and in fitting the rare-earth ions. However, for the nd^N ions it allows point groups 9 through 15 to be fitted with the same set of phenomenological B_{kq} (B_{20}, B_{40}, and ReB_{44}), and point groups 16 through 20 to be fitted with B_{20}, B_{40}, and ReB_{43}. Thus, for the computation of the energy levels, point groups 9 through 15 and 28 through 32 have B_{20}, B_{40}, and ReB_{44} (in the cubic groups $B_{20} = 0$, and B_{44} is related to B_{40}; see table 8.3). Similarly, point groups 16 through 27 have B_{20}, B_{40}, and ReB_{43} (with $B_{43} = 0$ for point groups 21 to 27; see table 8.2). These restrictions greatly reduce the computation.

8.2 Problems

1. Using the Hamiltonian in equation (8.2), prove the statement that

$$B_{k-q} = (-1)^q B^*_{kq}$$

by assuming that H_3 is real.

2. In the point group S_4, the tensors C_{1-1} and C_{11} form a basis for Γ_3 and Γ_4, respectively (table 25, Koster et al, 1963). From table 26 of

Koster et al we have $\Gamma_3 \times \Gamma_4 = \Gamma_1$. Show that the tensors formed by recoupling the product $C_{1-1}C_{11}$ using equation (1.27) form a basis for Γ_1.

3. Using the same methods as in problem 2, determine what bases are formed by $C_{10}C_{1-1}$ and $C_{10}C_{11}$ (C_{10} is a basis for Γ_2).

8.3 Annotated Bibliography and References

Ballhausen, C. J. (1962), Introduction to Ligand Field Theory, McGraw-Hill, New York, NY, chapter 4.

Hufner, S. (1978), Optical Spectra of Transparent Rare Earth Compounds, Academic Press, New York, NY. Chapter 3 has much material which is appropriate here.

Koster, G. F., J. O. Dimmock, R. G. Wheeler, and H. Statz (1963), Properties of the Thirty-Two Point Groups, MIT Press, Cambridge, MA.

Polo, S. R. (1961, June 1), Studies on Crystal Field Theory, Volume I--Text, Volume II--Tables, RCA Laboratories, under contract to Electronics Research Directorate, Air Force Cambridge Research Laboratories, Office of Aerospace Research, Contract No. AF 19(604)-5541. [Volume II gives date as June 1, 1961.]

Watanabe, H. (1966), Operator Methods in Ligand Field Theory, Prentice-Hall, Englewood Cliffs, NJ, chapter 4.

Wybourne, B. G. (1965), Spectroscopic Properties of Rare Earths, Wiley, New York, NY, chapter 6. (Many entries in table 6-1 are incorrect. For the correct relations, see C. A. Morrison and R. P. Leavitt, Handbook for the Physics and Chemistry of Rare Earths, 1982, North Holland Publishers, New York, NY, p 482.)

9. MATRIX ELEMENTS OF H_3 IN TOTAL ANGULAR MOMENTUM STATES FOR THE ELECTRONIC CONFIGURATION $n\ell^N$

9.1 Discussion

In order to make full use of tabulated data in our calculations, it is necessary to make some modifications in equation (8.1). Nielson and Koster (1963) have calculated the reduced matrix elements of the unit spherical tensors introduced by Racah. As was shown in equation (4.3), the $C_{kq}(i)$ can be written in terms of those tensors as

$$C_{kq}(i) = \langle \ell \| C_k \| \ell \rangle \, u_q^k(i) \tag{9.1}$$

and

$$\sum_i C_{kq}(i) = \langle \ell \| C_k \| \ell \rangle \, U_q^{(k)} \quad , \tag{9.2}$$

where

$$\langle \ell \| C_k \| \ell \rangle = \langle \ell(0)k(0) | \ell(0) \rangle \quad .$$

Thus equation (8.1) may be written

$$H_3 = \sum_{kq} B_{kq}^+ \langle \ell \| C_k \| \ell \rangle \, U_q^{(k)} \quad . \tag{9.3}$$

The matrix elements of H_3 in total orbital angular momentum states can now be written

$$\langle L'M_L'S'M_S'\alpha' | H_3 | LM_L SM_S\alpha \rangle \tag{9.4}$$

$$= \delta_{SS'}\delta_{M_SM_S'} \sum_{kq} B_{kq}^* \langle \ell \| C_k \| \ell \rangle \, \langle L(M_L)k(q) | L'(M_L') \rangle \, \langle L'S\alpha' \| U^{(k)} \| LS\alpha \rangle$$

where

$$\langle L'S\alpha' \| U^{(k)} \| LS\alpha \rangle = \left[\frac{2\ell+1}{2L'+1}\right]^{1/2} \left(L'S\alpha' \| U^{(k)} \| LS\alpha \right) \tag{9.5}$$

and the last expression in parentheses is tabulated in Nielson and Koster (1963).

Alternatively, we could use the tables of Polo (1961) to obtain the matrix elements of H_3 as given by

$$\langle L'M_L'S'M_S'\alpha' | H_3 | LM_LS\alpha \rangle$$

$$= \delta_{SS'}\delta_{M_SM_S'} \sum_{kq} B_{kq}^*(-1)^{L'-M_L} \begin{pmatrix} L' & k & L \\ -M_L' & q & M_L \end{pmatrix} \left(L'S\alpha' \| C^{(k)} \| LS\alpha \right) \tag{9.6}$$

where

$$\begin{pmatrix} L' & k & L \\ -M_L' & q & M_L \end{pmatrix}$$

is a 3j symbol and

$$C_q^{(k)} = \sum_{L'=1}^{N} C_{kq}(i) \quad .$$

The result given in equation (9.6) is much more convenient to use for computation than is equation (9.4).

The spin-orbit energy, H_2, can be written as

$$H_2 = \zeta \sqrt{s(s+1)\ell(\ell+1)} \sum_{\lambda} (-1)^\lambda V_{\lambda \ -\lambda}^{1 \ 1} \quad . \tag{9.7}$$

The matrix elements in total orbital angular momentum states are

$$\langle L'M_L'S'M_S'\alpha' | H_2 | LM_LSM_S\alpha \rangle = \zeta \sqrt{s(s+1)\ell(\ell+1)} (-1)^{M_L'-M_L} \langle L(M_L)1(\lambda) | L'(M_L') \rangle$$

$$\times \langle S(M_S)1(-\lambda) | S'(M_S') \rangle \langle L'S'\alpha' \| V^{11} \| LS\alpha \rangle$$

where

$$\langle L'S'\alpha'|V^{11}|LS\alpha\rangle = \frac{2}{\sqrt{3}}\left[\frac{2\ell+1}{(2L'+1)(2S'+1)}\right]^{1/2}(L'S'\alpha|V^{11}|LS\alpha) \quad , \qquad (9.8)$$

and the reduced matrix element in parentheses is tabulated by Nielson and Koster (1963).

The matrix elements of H_3 in total angular momentum states J ($\vec{J} = \vec{L}+\vec{S}$) can be written

$$\langle J'M'S\alpha'|H_3|JMLS\alpha\rangle$$

$$\qquad\qquad (9.9)$$

$$= \sum_{kq} B^*_{kq} \langle \ell|C_k|\ell\rangle \langle J(M)k(q)|J'(M')\rangle \langle J'L'S\alpha'|U^k|JLS\alpha\rangle$$

where

$$\langle J'L'S\alpha'|U^k|JLS\alpha\rangle = (-1)^{L-L'+J'-J}\sqrt{2J+1}\ W(kLJ'S;L'J)(L'S\alpha'|U^k|LS\alpha) \quad (9.10)$$

and again the reduced matrix element in parentheses is tabulated by Nielson and Koster (1963). Also, the quantity $\langle\ell|C_k|\ell\rangle$ in equation (9.9) is

$$\langle\ell|C_k|\ell\rangle = \langle\ell(0)k(0)|\ell(0)\rangle \quad . \qquad (9.11)$$

The matrix elements of H_2 in total angular momentum states, J, are given by

$$\langle J'M'L'S'\alpha'|H_2|JMLS\alpha\rangle$$

$$\qquad\qquad (9.12)$$

$$= -\zeta\ \sqrt{\ell(\ell+1)(2\ell+1)}\ \ W(S1J'L';S'L)(L'S'\alpha'|V^{11}|LS\alpha)\delta_{JJ'}\delta_{MM'} \quad ,$$

where the reduced matrix elements in LS space are tabulated by Nielson and Koster (1963).

9.2 Bibliography and References

International Tables for X-Ray Crystallography, Volume I (1952).

Judd, B. R. (1963), Operator Techniques in Atomic Spectroscopy, McGraw-Hill, New York, NY.

Koster, G. F., J. O. Dimmock, R. G. Wheeler, and H. Statz (1963), Properties of the Thirty-Two Point Groups, MIT Press, Cambridge, MA.

Nielson, C. W., and G. F. Koster (1963), Spectroscopic Coefficients for p^n, d^n, and f^n Configurations, MIT Press, Cambridge, MA.

Polo, S. R. (1961 June 1), Studies on Crystal Field Theory, Volume I--Text, Volume II--Tables, RCA Laboratories, under contract to Electronics Research Directorate, Air Force Cambridge Research Laboratories, Office of Aerospace Research, contract No. AF 19(604)-5541.

Racah, G. (1942), Theory of Complex Spectra--I, Phys. Rev. 61, 186; II, Phys. Rev. 62 (1942), 438; III, Phys. Rev. 63 (1942), 367; IV, Phys. Rev. 76 (1949), 1352.

Rotenberg, M., R. Bevins, N. Metropolis, and J. K. Wooten, Jr. (1969), The 3-j and 6-j Symbols, MIT Press, Cambridge, MA.

10. GROUP THEORETICAL CONSIDERATIONS

10.1 Discussion

We do not go into the details of group theory here but discuss the use of the tables presented in numerous texts on group theory. An excellent text for physicists is Tinkham (1964). For our discussion here, as in section 8, we use Koster et al (1963). This reference uses the Bethe notation for the irreducible representations (Γ_i); the relation of the Bethe notation to the Mulligan notation (A_i, B_i, etc) is given in the appendices of Griffith (1964). In our discussion we use the Bethe notation for all the single groups and both notations for the cubic O group. In the double groups we use only the Bethe notation.

As our first example we consider a single d electron in a crystal field of S_4 symmetry. The crystal-field interaction can be obtained from table 8.1; it is

$$H_3 = B_{20}C_{20} + B_{32}^* C_{32} + B_{32}C_{3-2} + B_{40}C_{40} + B_{44}\left(C_{44} + C_{4-4}\right) \quad . \tag{10.1}$$

If the spin-orbit interaction is small and the other configurations remote, we can at present ignore the B_{32} term and assume that angular wave functions are Y_{2m} ($|2m\rangle$), with $-2 \leq m < 2$. With these assumptions we use table 30 of Koster et al (1963) to find that

$$D_2^+ \rightarrow \Gamma_1 + 2\Gamma_2 + \Gamma_3 + \Gamma_4 \quad , \tag{10.2}$$

where we have used the D^+ table since $I|2m> = (-1)^2|2m>$; that is, the d electrons have positive parity. The wave functions for nd^N also have positive parity since $I|LSd^N> = (-1)^{2N}|LSd^N>$. But for the f electrons, $I|LSf^N> = (-1)^{3N}|LSf^N>$, which is even for even N and odd for odd N. Thus, for a single f electron we would use the D_3^- entry in table 30 of Koster et al. The result given in equation (10.2) shows that the five d electron states, which are degenerate in the free ion, would be split into four levels in the presence of the crystal field; these levels are shown in figure 10.1. All the free-ion degeneracy is removed except for the Γ_3,Γ_4 degeneracy. We can detect this degeneracy in character table 25 of Koster et al (1963, p 50) by observing that the product of the characters for Γ_3 and Γ_4 gives unity for all the group operations of S_4. We can also check this product by directly taking the matrix elements of H_3 using wave functions that transform as Γ_3 or Γ_4 in S_4. To find the wave functions which transform according to the irreducible representation, we use the operation S_4 on the states $|2m>$. That is,

$$S_4|2m> = IC_4^{-1}|2m>$$

$$(10.3)$$

$$= e^{-i\pi m/2}|2m>$$

and for $m = 0$

$$S_4|20> = |20> \quad .$$

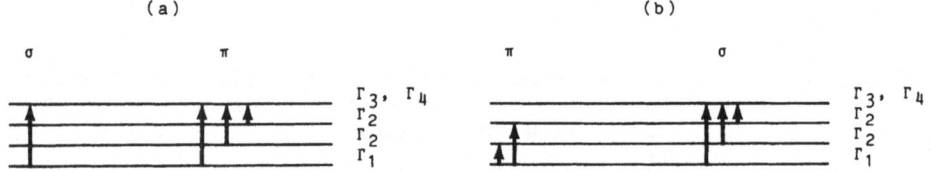

Figure 10.1. Hypothetical splitting of a single d electron in a crystal field of S_4 symmetry: (a) magnetic dipole and (b) electric dipole.

That is, the character is 1, so that $|20\rangle$ transforms as Γ_1 (see table 25 of Koster et al). Also, since

$$S_4|2\pm2\rangle = -|2\pm2\rangle \quad ,$$

the character is -1. From table 25 under the operation S_4, we find that the states with character -1 are Γ_2, so that the two wave functions $|22\rangle$ and $|2-2\rangle$ transform as Γ_2. Similarly,

$$S_4|2\pm1\rangle = \mp i|2\pm1\rangle \quad ;$$

from table 25 we see that $|21\rangle$ transforms as Γ_3 and $|2-1\rangle$ as Γ_4. It is not difficult to show that

$$\langle21|H_3|21\rangle = \langle2-1|H_3|2-1\rangle \quad , \tag{10.4}$$

which shows directly that the energy levels for Γ_3 and Γ_4 are degenerate.

The matrix elements $\langle2m'|H_3|2m\rangle$ of the crystal field are given by table 10.1, where the results given for Γ_2 are obtained from equation (3.18). From the results of table 10.1, we obtain the energy levels given in table 10.2 for Ti^{3+} for a particular choice of B_{kq}. Table 10.2 also gives the energy levels when the spin-orbit interaction is included.

The decomposition of the free-ion state onto the S_4 symmetry given in equation (10.2) gives the dimensions of the secular determinant which has to be solved to determine the energy levels of the system. In the case considered in table 10.1, the dimensions of the secular determinants are 1 for Γ_1, Γ_3, and Γ_4, and 2×2 for Γ_2. If for example we had an H state in S_4 symmetry (L = 5) of d^N, then, from table 30 of Koster et al (1963), we have

$$D_5^+ = 3\Gamma_1 + 4\Gamma_2 + 3\Gamma_3 + 3\Gamma_4 \tag{10.5}$$

and we would have to solve a 3×3 determinant for Γ_1, a 4×4 determinant for Γ_2, and a 3×3 determinant for Γ_3 or Γ_4.

TABLE 10.1. MATRIX ELEMENTS OF CRYSTAL FIELD FOR A SINGLE d ELECTRON IN S_4 SYMMETRY

Irreducible representation	Energy
Γ_1	$-\dfrac{2}{7} B_{20} + \dfrac{2}{7} B_{40}$
Γ_2^+	$-\dfrac{2}{7} B_{20} + \dfrac{1}{21} B_{40} + \dfrac{\sqrt{70}}{21} B_{44}$
Γ_2^-	$-\dfrac{2}{7} B_{20} + \dfrac{1}{21} B_{40} - \dfrac{\sqrt{70}}{21} B_{44}$
$\Gamma_{3,4}$	$\dfrac{1}{7} B_{20} - \dfrac{4}{21} B_{40}$

TABLE 10.2. ENERGY LEVELS FOR Ti^{3+} (3d) IN S_4 SYMMETRY[a]

$\zeta = 0$		$\zeta = 158$ cm^{-1}	
IR	E (cm^{-1})	IR	E (cm^{-1})
Γ_1	0	$\Gamma_{5,6}$	0
Γ_2	395	$\Gamma_{5,6}$	0
Γ_2	2931	$\Gamma_{7,8}$	2938
$\Gamma_{3,4}$	3720	$\Gamma_{5,6}$	3662
		$\Gamma_{7,8}$	3827

[a] $B_{20} = 394.7$ cm^{-1}, $B_{40} = -7932$ cm^{-1}, $B_{44} = 3182$ cm^{-1}. These are approximate crystal-field parameters for Ti^{3+} in the Ga site of $Gd_3Sc_2Ga_3O_{12}$, gadolinium scandium gallium garnet.
IR = irreducible representation
E = energy

If we wish to determine the energy levels experimentally, it is informative to investigate the use of polarized radiation. The magnetic dipole operator is

$$H_{md} = \mu_\beta \vec{B} \cdot (\vec{L} + 2\vec{S})$$

$$= \mu_\beta \sum_\alpha B_\alpha^* (L_\alpha + 2S_\alpha) \quad , \tag{10.6}$$

where B is the strength of the magnetic field and μ_β is the Bohr magneton. From group theory, we can deduce the transitions induced by equation (10.6) by using first table 25 of Koster et al. This table shows that L_z and S_z are basis functions for Γ_1; L_{+1} and S_{+1} are basis functions for Γ_3; and L_{-1} and S_{-1} are basis functions for Γ_4. Then from table 26 of Koster et al, we have, for $B \| Z$ ($L_z \rightarrow \Gamma_1$),

$$\Gamma_1 \times \Gamma_1 = \Gamma_1 \quad ,$$

$$\Gamma_1 \times \Gamma_2 = \Gamma_2 \quad ,$$

$$\Gamma_1 \times \Gamma_3 = \Gamma_3 \quad ,$$

$$\Gamma_1 \times \Gamma_4 = \Gamma_4 \quad ,$$

96

and for $B|X$ or $B|Y$ $(L_\pm \to \Gamma_3, \Gamma_4)$,

$$\Gamma_3 \times \Gamma_1 = \Gamma_3 \quad , \qquad \Gamma_4 \times \Gamma_1 = \Gamma_4 \quad ,$$

$$\Gamma_3 \times \Gamma_2 = \Gamma_4 \quad , \qquad \Gamma_4 \times \Gamma_2 = \Gamma_3 \quad ,$$

$$\Gamma_3 \times \Gamma_3 = \Gamma_2 \quad , \qquad \Gamma_4 \times \Gamma_3 = \Gamma_1 \quad ,$$

$$\Gamma_3 \times \Gamma_4 = \Gamma_1 \quad , \qquad \Gamma_4 \times \Gamma_4 = \Gamma_2 \quad .$$

The allowed transitions (in absorption) are shown in figure 10.1a, where the labeling is according to the orientation of the light polarizer. That is, the polarization is determined by the orientation of the electric vector; for π polarization, $E|Z$ (and $B\perp Z$), and for σ polarization, $E\perp Z$ (and $B|Z$).

The electric dipole operator is

$$H'_{ed} = e\vec{E} \cdot \vec{r}$$

$$= e \sum_\alpha E_\alpha^* r C_{1\alpha}(\hat{r}) \quad . \tag{10.7}$$

From table 25 (Koster et al, 1963), we find that $C_{10}(z)$ is a basis for Γ_2 while $C_{1\pm 1}$ (or $x \pm iy$) has the same representations as $L_{\mp 1}$. For C_{10}, from table 26 (Koster et al), we have

$$\Gamma_2 \times \Gamma_1 = \Gamma_2 \quad ,$$

$$\Gamma_2 \times \Gamma_2 = \Gamma_1 \quad ,$$

$$\Gamma_2 \times \Gamma_3 = \Gamma_4 \quad ,$$

$$\Gamma_2 \times \Gamma_4 = \Gamma_3 \quad ,$$

and the allowed transitions (in absorption) are shown in figure 10.1b. In the electric dipole case we have to assume that the odd terms in the crystal field given in equation (10.1) mix either the p or f configuration; otherwise, the electric dipole matrix elements vanish.

As a second example we consider the d^3 configuration in the crystal field given in equation (10.1). The Hund ground state is ^4F ($S = 3/2$, $L = 3$) and the only other state with $S = 3/2$ is the ^4P (all the states of p^N, d^N, and f^N are given on pages 1 through 3 in Nielson and Koster, 1963, and pages 15-14 through 15-20 in Polo, 1961). In the absence of spin-orbit coupling, the

^{4}p state is the only state that couples to the ^{4}F Hund ground state. From table 30 of Koster et al we have

$$D_3^+ \rightarrow \Gamma_1 + 2\Gamma_2 + 2\Gamma_{3,4} \quad ,$$

$$\text{(10.8)}$$

$$D_1^+ \rightarrow \Gamma_1 + \Gamma_{3,4} \quad ,$$

and we have a 2×2 secular equation for Γ_1 and Γ_2, and a 3×3 secular equation for the $\Gamma_{3,4}$ states. Operating on the states $|LM\rangle$ with the operator S_4 produces the values of M belonging to the different irreducible representations. The resulting states are

$$|30\rangle \; , \; |10\rangle \; , \qquad \qquad \text{for} \;\; \Gamma_1 \;\; ,$$

$$|3\text{-}2\rangle \; , \; |32\rangle \; , \qquad \qquad \text{for} \;\; \Gamma_2 \;\; , \qquad \qquad \text{(10.9)}$$

$$|3\text{-}1\rangle \; , \; |33\rangle \; , \;\; |1\text{-}1\rangle \;\; , \quad \text{for} \;\; \Gamma_4 \;\; .$$

The states for Γ_3 are obtained by changing the projections (M → -M) of the Γ_4 states. The matrix elements of the crystal field for the states given in equation (10.9) are presented in table 10.3. The results given in table 10.3 are also applicable to the configuration d^7, provided that all values are multiplied by minus one. This latter result is obtained by the rules for the states conjugate to the ℓ^N configuration, which has $4\ell+2-N$ electrons; these rules are given in Nielson and Koster (1963). The results given in table 10.3 were taken from the Polo tables, and equation (9.6) was used. The 3-j symbols were found in Rotenberg et al (1969).

Before the energy levels can be found using the results of table 10.3, the Coulomb interaction must be added to the diagonal elements. From Nielson and Koster, these are

$$H_1\left(^4F\right) = -\frac{15}{49} F^{(2)} - \frac{72}{441} F^{(4)} \quad ,$$

$$\text{(10.10)}$$

$$H_1\left(^4P\right) = -\frac{147}{441} F^{(4)} \quad .$$

The energy levels of d^3 in S_4 symmetry is given in table 10.4 for a representative set of B_{KQ} and $F^{(k)}$. Included in the table is the effect of spin-orbit coupling, which we consider next.

The above examples have been restricted to the orbital angular momentum only. This restricts the use of group theory to the single group.

TABLE 10.3. MATRIX ELEMENTS, IN S_4 SYMMETRY, OF CRYSTAL FIELD FOR ^4F AND ^4P STATES OF nd^3

Note: For the corresponding matrix elements for ^4F and ^4P states of nd^7, multiply all entries by -1.[a]

L'M'	L M	IR[b]	B_{20}	B_{40}	B_{44}
3 0	3 0	Γ_1	-4/35	2/7	0
3 0	1 0	Γ_1	12/35	-4/21	0
1 0	1 0	Γ_1	2/5	0	0
3-2	3-2	Γ_2	0	-1/3	0
3-2	3 2	Γ_2	0	0	$\sqrt{70}/21$
3 2	3 2	Γ_2	0	-1/3	0
3-1	3-1	Γ_4	-3/35	1/21	0
3-1	3 3	Γ_4	0	0	$\sqrt{42}/21$
3-1	1-1	Γ_4	$2\sqrt{6}/35$	$\sqrt{6}/21$	0
3 3	3 3	Γ_4	1/7	1/7	0
3 3	1-1	Γ_4	0	0	$2\sqrt{7}/21$
1-1	1-1	Γ_4	-1/5	0	0

[a]The matrix elements of Γ_3 are not given, as they are equal to those for Γ_4. Before the matrix is diagonalized, the Coulomb energies from Nielson and Koster (1963) for the ^4F and ^4P states should be added to the diagonal elements.
[b]IR = irreducible representation.

TABLE 10.4. ENERGY LEVELS FOR Cr^{3+} ^4F AND ^4P LEVELS IN S_4 SYMMETRY[a]

Free ion	$\zeta = 0$		$\zeta = 275$ cm^{-1}	
	IR	Energy (cm^{-1})	IR	Energy (cm^{-1})
^4F	Γ_1	0	$\Gamma_{5,6}$	0
	$\Gamma_{3,4}$	352	$\Gamma_{7,8}$	94
	$\Gamma_{3,4}$	1,574	$\Gamma_{7,8}$	249
	Γ_2	2,203	$\Gamma_{7,8}$	496
	Γ_2	3,672	$\Gamma_{5,6}$	622
			$\Gamma_{5,6}$	788
			$\Gamma_{7,8}$	1,713
			$\Gamma_{7,8}$	1,748
			$\Gamma_{5,6}$	1,783
			$\Gamma_{5,6}$	1,901
			$\Gamma_{5,6}$	2,381
			$\Gamma_{7,8}$	2,460
			$\Gamma_{7,8}$	3,918
			$\Gamma_{5,6}$	3,930
^4P	$\Gamma_{3,4}$	16,294	$\Gamma_{5,6}$	16,286
	Γ_1	16,504	$\Gamma_{5,6}$	16,373
			$\Gamma_{7,8}$	16,442
			$\Gamma_{7,8}$	16,621
			$\Gamma_{7,8}$	16,741
			$\Gamma_{5,6}$	16,771

[a]$B_{20} = 296.9$ c^{-1}, $B_{40} = 4597$ cm^{-1}, $B_{44} = 1844$ cm^{-1}, $F^{(2)} = 74,201$ cm^{-1}, $F^{(4)} = 45,822$ cm^{-1}. The $F^{(2)}$, $F^{(4)}$, and ζ (275 cm^{-1}) values are for the free ion.
IR = irreducible representation.

If we consider the case of the d^3 as above but assume the spin-orbit energy to be strong, we use total angular momentum states, J, with $\vec{J} = \vec{L} + \vec{S}$. We further restrict the discussion to the states ^4F_J with $3/2 \leq J \leq 9/2$. Thus, we have the levels $^4F_{3/2}$, $^4F_{5/2}$, $^4F_{7/2}$, and $^4F_{9/2}$ to consider. From table 30 (Koster et al) we obtain

$$D_{3/2}^+ \rightarrow \Gamma_{5,6} + \Gamma_{7,8} \quad ,$$

$$D_{5/2}^+ \rightarrow \Gamma_{5,6} + 2\Gamma_{7,8} \quad ,$$

$$\text{(10.11)}$$

$$D_{7/2}^+ \rightarrow 2\Gamma_{5,6} + 2\Gamma_{7,8} \quad ,$$

$$D_{9/2}^+ \rightarrow 3\Gamma_{5,6} + 2\Gamma_{7,8} \quad ,$$

and we see that the secular equation for Γ_5 or Γ_6 is 7×7, and for Γ_7 or Γ_8 the secular equation is also 7×7. We see that the inclusion of the spin-orbit energy greatly increases the difficulty of the problem. The wave functions belonging to the different irreducible representations can be obtained from table 25 of Koster et al and are

$$|J \tfrac{1}{2} + 4p\rangle \quad , \qquad \text{where} \quad p = 0, \pm 1, \ldots, \quad |\tfrac{1}{2} + 4p| \leq J, \quad \text{for} \quad \Gamma_5 \quad ,$$

$$\text{(10.12)}$$

$$|J - \tfrac{3}{2} + 4p\rangle \quad , \qquad \text{where} \quad p = 0, \pm 1, \ldots, \quad |-\tfrac{3}{2} + 4p| \leq J, \quad \text{for} \quad \Gamma_7 \quad .$$

The resulting energy levels of the Cr^{3+} ion with the Slater parameters and the spin-orbit constant from the free ion are given in table 10.4. The crystal-field parameters are rough estimates for the crystal field for Cr^{3+} in the Ga site (S_4) in the material $Gd_3Sc_2Ga_3O_{12}$.

The previous two examples were for the S_4 group and were simple to manipulate in that the group was cyclic (all the operations can be expressed in terms of a single generator, S_4).

We consider a single d electron in a crystal field of D_2 symmetry. The crystal-field interaction, given by use of table 8.1, is

$$H_3 = B_{20}C_{20} + B_{22}(C_{22} + C_{2-2}) + ImB_{32}(C_{32} - C_{3-2})$$

$$\text{(10.13)}$$

$$+ B_{40}C_{40} + B_{42}(C_{42} + C_{4-2}) + B_{44}(C_{44} + C_{4-4}) \quad .$$

The operations of D_2 are given in table 17 of Koster et al and are $C_2 = C_2(z)$, $C_2' = C_2(y)$, and $C_2'' = C_2(x)$. The result of these operations on $|\ell m\rangle$ is given in problem 2 of chapter 1. These are

$$C_2(y)|\ell m\rangle = (-1)^{\ell+m}|\ell - m\rangle \quad ,$$

$$C_2(z)|\ell m\rangle = (-1)^m |\ell m\rangle \quad , \qquad (10.14)$$

$$C_2(x)|\ell m\rangle = (-1)^\ell |\ell - m\rangle \quad .$$

Now from table 21 of Koster et al, the $|2m\rangle$ states decompose to

$$D_2^+ \to 2\Gamma_1 + \Gamma_2 + \Gamma_3 + \Gamma_4 \quad , \qquad (10.15)$$

and the Γ_1 states are

$$\psi_1 = |20\rangle \quad , \quad \frac{1}{\sqrt{2}}[|22\rangle + |2-2\rangle] \qquad (10.16)$$

which are essentially the crystal potential for the C_{2q} terms in equation (10.13).

For the other representation we take the combination

$$\psi_2 = \frac{1}{\sqrt{2}}[|21\rangle - |2-1\rangle] \quad \text{for } \Gamma_2 \quad , \qquad (10.17)$$

$$\psi_3 = \frac{1}{\sqrt{2}}[|22\rangle - |2-2\rangle] \quad \text{for } \Gamma_3 \quad , \qquad (10.18)$$

$$\psi_4 = \frac{1}{\sqrt{2}}[|21\rangle + |2-1\rangle] \quad \text{for } \Gamma_4 \quad , \qquad (10.19)$$

which can be verified by using equation (10.14) and the character table for the D_2 group in Koster et al.

10.2 Annotated Bibliography and References

Ballhausen, C. J. (1962), Introduction to Ligand Field Theory, McGraw-Hill, New York, NY.

Griffith, J. S. (1961), The Theory of Transition-Metal Ions, Cambridge University Press, Cambridge, U. K.

Konig, E., and S. Kremer (1977), Ligand Field Energy Diagrams, Plenum Press, New York, NY. This book contains a large number of energy levels of $3d^N$ ions in different crystalline environments. Pages 18 through 22 describe the different notation used to specify the constants (crystal-field parameters) of the system.

Koster, G. F., J. O. Dimmock, R. G. Wheeler, and H. Statz (1963), Properties of the Thirty-Two Point Groups, MIT Press, Cambridge, MA.

McClure, D. S. (1959), Electronic Spectra of Molecules and Ions in Crystals, Part II.--Spectra of Ions in Crystals, Solid State Phys. $\underline{9}$, 399.

Nielson, C. W., and G. F. Koster (1963), Spectroscopic Coefficients for the p^n, d^n, and f^n Configurations, MIT Press, Cambridge, MA.

Polo, S. R. (1961, June 1), Studies on Crystal Field Theory, Volume I--Text, Volume II--Tables, RCA Laboratories, under contract to Electronics Research Directorate, Air Force Cambridge Research Laboratories, Office of Aerospace Research, contract No. AF 19(604)-5541. [Volume II gives date as June 1, 1961.]

Rotenberg, M., R. Bevins, N. Metropolis, and J. K. Wooten, Jr. (1969), The 3-j and 6-j Symbols, MIT Press, Cambridge, MA.

Tinkham, M. (1964), Group Theory and Quantum Mechanics, McGraw-Hill, New York, NY.

11.   NUMERICAL EXAMPLE: ^4F_J STATES OF Nd^{3+} ($4f^3$)

As a numerical example of the calculation of the crystal-field splitting, we calculate the splitting for Nd^{3+} ($4f^3$) in a field of S_4 point symmetry. We assume that the levels are pure ^4F_J. We assume that L, S, and J are all good quantum numbers; then we consider matrix elements of H_3 in equation (9.9) with J' = J and L' = L. Thus,

$$\langle JM'LS|H_3|JMLS\rangle = \sum_{kq} B^*_{kq} \langle J(M)k(q)|J(M')\rangle \langle J\|U^k\|J\rangle \langle \ell(0)k(0)|\ell(0)\rangle \quad . \quad (11.1)$$

The values of the reduced matrix elements $\langle J\|U^k\|J\rangle \langle \ell\|C_k\|\ell\rangle$ for the ^4F state of Nd^{3+} are as in table 11.1. In obtaining these values we have used Nielson and Koster's results (1963) in equations (9.9) and (9.10) for the reduced matrix elements, $(LS\|U^k\|LS)$, for L = 3 and S = 3/2; the Racah coefficients (6-j symbols) are found in Rotenberg et al (1969).

The calculation of the energy levels is made somewhat simpler by using wave functions that transform according to some irreducible representation of the S_4 group (Koster et al, 1963). The irreducible representations of the S_4 group are all one-dimensional, but, since the ion under investigation has an odd number of electrons, the energy levels will be at least doubly degenerate. Thus, of the four irreducible representations--Γ_5, Γ_6, Γ_7, and Γ_8--only two need be chosen; we chose Γ_5 and Γ_7. The energy corresponding to Γ_6 is degenerate with Γ_5, and that corresponding to Γ_8 is degenerate with Γ_7. The wave functions belonging to Γ_7 with a particular J value are

$$\left| J\, \frac{8k+1}{2} \right\rangle \, , \qquad -\frac{2J+1}{8} \leq k \leq \frac{2J-1}{8} \, ,$$

J	$k=2$	$k=4$	$k=6$
$\frac{3}{2}$	$\frac{2}{5}\left[\frac{1}{5}\right]^{1/2}$	0	0
$\frac{5}{2}$	$\frac{11}{2\cdot5}\left[\frac{1}{2\cdot5\cdot7}\right]^{1/2}$	$\frac{1}{2}\left[\frac{1}{2\cdot3\cdot7}\right]^{1/2}$	0
$\frac{7}{2}$	$\frac{1}{3}\left[\frac{5}{3\cdot7}\right]^{1/2}$	$\frac{1}{2\cdot3}\left[\frac{1}{7\cdot11}\right]^{1/2}$	$-\frac{2\cdot5}{3}\left[\frac{1}{3\cdot11\cdot13}\right]^{1/2}$
$\frac{9}{2}$	$\frac{1}{2\cdot3}\left[\frac{11}{2\cdot3}\right]^{1/2}$	$-\frac{1}{2\cdot3}\left[\frac{13}{2\cdot11}\right]^{1/2}$	$\frac{5}{3}\left[\frac{1}{2\cdot3\cdot11\cdot13}\right]^{1/2}$

^{a}Note: $2\cdot5 = 10$, etc.

and those for Γ_5 are

$$\left| J\;\frac{8k-3}{2}\right\rangle , \qquad -\frac{2J-3}{8}\leq k\leq\frac{2J+3}{8} ,$$

where k is an integer, and the number of k values occurring for a given J is the number of times a representation will occur.   The number of B_{kq} for the calcium site (S_4) in calcium tungstate is five: B_{20}, B_{40}, B_{60}, B_{44}, and B_{64}.   Of these parameters only B_{64} is complex.   The matrix elements of the crystal field given in the above equation are presented explicitly under particular states in the following paragraphs.

11.1 $^4F_{3/2}$

This level of the free ion is split into two doublets by the crystalline field.   The wave functions corresponding to Γ_7 and Γ_5 are $|\frac{3}{2}\,\frac{1}{2}\rangle$ and $|\frac{3}{2}-\frac{3}{2}\rangle$ respectively.   From equation (11.1) we have

$$\left\langle\frac{3}{2}\,\frac{1}{2}\right|H_3\left|\frac{3}{2}\,\frac{1}{2}\right\rangle = -\frac{2}{25}B_{20} = E\left(\Gamma_7\,\frac{3}{2}\right) \tag{11.2}$$

and

$$\left\langle\frac{3}{2}-\frac{3}{2}\right|H_3\left|\frac{3}{2}-\frac{3}{2}\right\rangle = \frac{2}{25}B_{20} = E\left(\Gamma_5\,\frac{3}{2}\right) , \tag{11.3}$$

where the appropriate values of the reduced matrix elements in equations (11.2) and (11.3) were taken from table 11.1. The total splitting of the $^4F_{3/2}$ state is then

$$\frac{4}{25} B_{20} \ . \tag{11.4}$$

11.2 $^4F_{5/2}$

Unlike the previous case, this state contains two Γ_5's, and their wave functions are $|\frac{5}{2} \frac{5}{2} >$ and $|\frac{5}{2} - \frac{3}{2} >$. The wave function for the Γ_7 state is $|\frac{5}{2} \frac{1}{2} >$. The energy for Γ_7 is

$$< \frac{5}{2} \frac{1}{2} |H_3| \frac{5}{2} \frac{1}{2} > = \frac{11}{700} \left[-4B_{20} + \frac{50}{33} B_{40} \right] = E\left(\Gamma_7 \frac{5}{2}\right) \ . \tag{11.5}$$

The necessary matrix elements for the energy in Γ_5 are

$$< \frac{5}{2} \frac{5}{2} |H_3| \frac{5}{2} \frac{5}{2} > = \frac{11}{700} \left[5B_{20} + \frac{25}{38} B_{40} \right] = b_{11} \ , \tag{11.6}$$

$$< \frac{5}{2} - \frac{3}{2} |H_3| \frac{5}{2} - \frac{3}{2} > = \frac{11}{700} \left[-B_{20} - \frac{25}{11} B_{40} \right] = b_{22} \ , \tag{11.7}$$

$$< \frac{5}{2} - \frac{3}{2} |H_3| \frac{5}{2} \frac{5}{2} > = \frac{1}{6\sqrt{14}} B_{44} = b_{12} \ . \tag{11.8}$$

The two energy levels corresponding to Γ_5 are

$$E_1\left(\Gamma_5 \frac{5}{2}\right) = \frac{b_{11} + b_{22} + \left[(b_{11} - b_{22})^2 + 4b_{12}b_{12}^* \right]^{1/2}}{2} \ , \tag{11.9}$$

$$E_2\left(\Gamma_5 \frac{5}{2}\right) = \frac{b_{11} + b_{22} - \left[(b_{11} - b_{22})^2 + 4b_{12}b_{12}^* \right]^{1/2}}{2} \ . \tag{11.10}$$

11.3 $^4F_{7/2}$

This state contains two Γ_5's and two Γ_7's. The matrix elements for Γ_7 are

$$\langle \tfrac{7}{2} \tfrac{1}{2} | H_3 | \tfrac{7}{2} \tfrac{1}{2} \rangle = \frac{1}{99} \left[-\frac{55}{7} B_{20} + \frac{9}{14} B_{40} + \frac{50}{13} B_{60} \right] = a_{11} \quad , \quad (11.11)$$

$$\langle \tfrac{7}{2} -\tfrac{7}{2} | H_3 | \tfrac{7}{2} -\tfrac{7}{2} \rangle = \frac{1}{99} \left[11 B_{20} + \frac{1}{2} B_{40} - \frac{10}{13} B_{60} \right] = a_{22} \quad , \quad (11.12)$$

$$\langle \tfrac{7}{2} -\tfrac{7}{2} | H_3 | \tfrac{7}{2} \tfrac{1}{2} \rangle = \frac{1}{99} \left[\frac{1}{\sqrt{2}} B_{44} - \frac{30}{13} \sqrt{10} \, B_{64} \right] = a_{12} \quad . \quad (11.13)$$

The two energy levels are

$$E_1\left(\Gamma_7 \tfrac{7}{2}\right) = \frac{a_{11} + a_{22} + \left[(a_{11} - a_{22})^2 + 4 a_{12} a_{12}^* \right]^{1/2}}{2} \quad , \quad (11.14)$$

$$E_2\left(\Gamma_7 \tfrac{7}{2}\right) = \frac{a_{11} + a_{22} - \left[(a_{11} - a_{22})^2 + 4 a_{12} a_{12}^* \right]^{1/2}}{2} \quad . \quad (11.15)$$

The matrix elements for Γ_5 are*

$$\langle \tfrac{7}{2} \tfrac{5}{2} | H_3 | \tfrac{7}{2} \tfrac{5}{2} \rangle = \frac{1}{99} \left[\frac{11}{7} B_{20} - \frac{13}{14} B_{40} + \frac{50}{13} B_{60} \right] = b_{11} \quad , \quad (11.16)$$

$$\langle \tfrac{7}{2} -\tfrac{3}{2} | H_3 | \tfrac{7}{2} -\tfrac{3}{2} \rangle = \frac{1}{99} \left[-\frac{33}{7} B_{20} - \frac{3}{14} B_{40} - \frac{90}{13} B_{60} \right] = b_{22} \quad , \quad (11.17)$$

$$\langle \tfrac{7}{2} -\tfrac{3}{2} | H_3 | \tfrac{7}{2} \tfrac{5}{2} \rangle = \frac{1}{99} \left[\sqrt{15/14} \, B_{44} + \frac{20}{13} \sqrt{21/2} \, B_{64} \right] = b_{12} \quad . \quad (11.18)$$

The corresponding energies are given by substituting the above values of b_{ij} into equations (11.9) and (11.10).

---

*The symbol a_{ij} will be used for the matrix elements in Γ_7 and b_{ij} for those in Γ_5 to avoid introducing new symbols for each new value of J.

11.4 $^4F_{9/2}$

The number of Γ_7's in this state is three, with two Γ_5's. The matrix elements of the crystal field for Γ_7 are

$$\langle \tfrac{9}{2}\ \tfrac{9}{2}\ |H_3|\ \tfrac{9}{2}\ \tfrac{9}{2} \rangle = \frac{7}{396} \left[6B_{20} - \frac{18}{7} B_{40} + \frac{30}{91} B_{60} \right] = a_{11}\ , \qquad (11.19)$$

$$\langle \tfrac{9}{2}\ \tfrac{1}{2}\ |H_3|\ \tfrac{9}{2}\ \tfrac{1}{2} \rangle = \frac{7}{396} \left[-4B_{20} - \frac{18}{7} B_{40} - \frac{80}{91} B_{60} \right] = a_{22}\ , \qquad (11.20)$$

$$\langle \tfrac{9}{2}\ -\tfrac{7}{2}\ |H_3|\ \tfrac{9}{2}\ -\tfrac{7}{2} \rangle = \frac{7}{396} \left[2B_{20} + \frac{22}{7} B_{40} - \frac{110}{91} B_{60} \right] = a_{33}\ , \qquad (11.21)$$

$$\langle \tfrac{9}{2}\ \tfrac{1}{2}\ |H_3|\ \tfrac{9}{2}\ \tfrac{9}{2} \rangle = \frac{7}{396} \left[-\frac{6}{7} \sqrt{5}\ B_{44} + \frac{150}{91} B_{64} \right] = a_{12}\ , \qquad (11.22)$$

$$\langle \tfrac{9}{2}\ -\tfrac{7}{2}\ |H_3|\ \tfrac{9}{2}\ \tfrac{1}{2} \rangle = \frac{7}{396} \left[-\frac{10}{7} \sqrt{5}\ B_{44} + \frac{30}{91} B_{64} \right] = a_{23}\ . \qquad (11.23)$$

The three energies are given by the solutions of

$$E^3 - (a_{11} + a_{22} + a_{33})E^2 + (a_{11}a_{22} + a_{11}a_{33} + a_{22}a_{33} - a_{23}a^*_{23} - a_{12}a^*_{12})E$$

$$+ a_{11}a_{23}a^*_{23} + a_{33}a_{12}a^*_{12} - a_{11}a_{22}a_{33} = 0\ .$$

The matrix elements for Γ_5 are

$$\langle \tfrac{9}{2}\ \tfrac{5}{2}\ |H_3|\ \tfrac{9}{2}\ \tfrac{5}{2} \rangle = \frac{7}{396} \left[B_{20} + \frac{17}{7} B_{40} + \frac{100}{91} B_{60} \right] = b_{11}\ , \qquad (11.25)$$

$$\langle \tfrac{9}{2}\ -\tfrac{3}{2}\ |H_3|\ \tfrac{9}{2}\ -\tfrac{3}{2} \rangle = \frac{7}{396} \left[-3B_{20} - \frac{3}{7} B_{40} + \frac{60}{91} B_{60} \right] = b_{22}\ , \qquad (11.26)$$

$$\langle \tfrac{9}{2}\ -\tfrac{3}{2}\ |H_3|\ \tfrac{9}{2}\ \tfrac{5}{2} \rangle = \frac{7}{396} \left[-\frac{5}{7} \sqrt{30}\ B_{44} - \frac{40}{91} \sqrt{6}\ B_{64} \right] = b_{12}\ . \qquad (11.27)$$

The energies $E_1\left(\Gamma_5\ \tfrac{9}{2}\right)$ and $E_2\left(\Gamma_5\ \tfrac{9}{2}\right)$ are given by equations (11.9) and (11.10), respectively, using the b_{ij} given above.

11.5 Calculations

Three of the crystal-field parameters can be obtained quite simply from the experimental data. These are B_{20}, B_{40}, and B_{60}. If we express the sums, S_i, in terms of the E_i where the E_i are experimental data, then we obtain

$$S_{J-1/2} = \sum_i E_i (\Gamma_7 J) \quad ,$$

and

$$S_1 = -\frac{2}{25} B_{20} \quad , \tag{11.28}$$

$$S_2 = -\frac{11}{175} B_{20} + \frac{1}{42} B_{40} \quad , \tag{11.29}$$

$$S_3 = \frac{2}{63} B_{20} + \frac{8}{693} B_{40} + \frac{40}{1287} B_{60} \quad , \tag{11.30}$$

$$S_4 = \frac{1}{9} B_{20} - \frac{7}{198} B_{40} - \frac{40}{1287} B_{60} \quad , \tag{11.31}$$

where S_4 is for the $J = 9/2$ level. These equations can be inverted to give

$$B_{20} = -\frac{25}{2} S_1 \quad , \tag{11.32}$$

$$B_{40} = -33 S_1 + 42 S_2 \quad , \tag{11.33}$$

$$B_{60} = \frac{1001}{40} S_1 - \frac{78}{5} S_2 + \frac{1287}{40} S_3 \quad . \tag{11.34}$$

The other crystal-field parameters are slightly more involved. From equations (11.9) and (11.10) we have

$$B_{44} = 3\sqrt{14} \left[W_1^2 - N_1^2 \right]^{1/2} \quad , \tag{11.35}$$

where

$$N_1 = -\frac{11}{4} S_1 + 2S_2 \quad \text{and} \quad W_1 = E_1 \left(\Gamma_5 \frac{5}{2} \right) - E_2 \left(\Gamma_5 \frac{5}{2} \right) \quad .$$

To determine B_{64}, we use equations (11.14) and (11.15) to give

$$a_{12} a_{12}^* = \frac{1}{4} \left[W_2^2 - N_2^2 \right] \tag{11.36}$$

108

where

$$N_2 = \frac{7}{6} S_1 - \frac{2}{3} S_2 + \frac{3}{2} S_3 \quad \text{and} \quad W_2 = E_1\left(\Gamma_7 \frac{7}{2}\right) - E_2\left(\Gamma_7 \frac{7}{2}\right) \ .$$

A similar expression can be obtained using equations (11.16) and (11.17) in equations (11.9) and (11.10), yielding

$$b_{12} b_{12}^* = \frac{1}{4} \left[W_3^2 - N_3^2 \right] \ , \tag{11.37}$$

where

$$N_3 = \frac{13}{6} S_1 - 2S_2 + \frac{7}{2} S_3 \quad \text{and} \quad W_3 = E_1\left(\Gamma_5 \frac{7}{2}\right) - E_2\left(\Gamma_5 \frac{7}{2}\right) \ .$$

Substituting equations (11.13) and (11.14) into the left side of equations (11.36) and (11.37), we obtain two equations for B_{64}. These two equations can be solved simultaneously for both real and imaginary parts of B_{64} to give

$$R_6 = \frac{13}{20} \left[\frac{3 \cdot 99}{4} \left(W_2^2 + W_3^2 - N_2^2 - N_3^2 \right) - 6\left(W_1^2 - N_1^2 \right) \right]^{1/2} \ , \tag{11.38}$$

$$\cos \theta = \frac{13}{20\sqrt{70}} \frac{\frac{15 \cdot 99}{8} \left(W_3^2 - N_3^2 \right) - 8\left(W_1^2 - N_1^2 \right) - \frac{7 \cdot 99}{8} \left(W_2^2 - N_2^2 \right)}{R_6\left(W_1^2 - N_1^2 \right)^{1/2}} \ , \tag{11.39}$$

where $B_{64} = R_6 e^{i\theta}$.

All the crystal-field parameters can be determined once the experimental data are taken on the $^4F_{3/2}$, $^4F_{5/2}$, and $^4F_{7/2}$ levels.

As tedious as the above procedures may have seemed, the crystal-field parameters we obtain are only approximate, since we have ignored L-S mixing by the spin-orbit coupling in the free ion and J mixing caused by the crystal field. Nevertheless, the crystal-field parameters obtained by the above procedure can serve as very good starting values in the fitting of a more sophisticated calculation to experimental crystal-field levels.

The crystal-field parameters B_{kq} obtained by the above procedure for Nd^{3+} in $CaWO_4$ are given as follows, along with crystal-field parameters for the same ion but with full diagonalization, that is, L-S mixing and J mixing (Wortman et al, 1977).

B_{kq}	B_{20}	B_{40}	B_{44}	B_{60}	B_{64}	
					Re	Im
Calculated above	403	-635	711	-219	885	0
Calculated with full diagon- alization	509	-866	1042	-509	903	243

11.6 References

Koster, G. F., J. O. Dimmock, R. G. Wheeler, and H. Statz (1963), Properties of the Thirty-Two Point Groups, MIT Press, Cambridge, MA.

Nielson, C. W., and G. F. Koster (1963), Spectroscopic Coefficients for the p^n, d^n, and f^n Configurations, MIT Press, Cambridge, MA.

Rotenberg, M., R. Bevins, N. Metropolis, and J. K. Wooten, Jr. (1969), The 3-j and 6-j Symbols, MIT Press, Cambridge, MA.

Wortman, D. E., C. A. Morrison, and N. Karayianis (1977, June), Rare Earth Ion-Host Lattice Interactions: 5.--Lanthanides in $CaWO_4$, Harry Diamond Laboratories, HDL-TR-1794.

12. CLASSICAL POINT-CHARGE MODEL

12.1 Discussion

In the simplest model of the crystal field, the point-charge model introduced by Bethe (1929), the lattice is replaced by an array of point charges placed at the nuclei of the constituent ions. A multipole expansion is made of the point-charge potential energy at the rare-earth site. Thus, if $\vec{R}_{\ell mn}(j)$ is the vector position of constituent j at site j in the ℓ, m, nth cell, we have

$$H_3 = \sum_{\ell mn} \sum_j \frac{-e^2 Z_j}{|\vec{R}_{\ell mn}(j) - \vec{r}|} \quad , \tag{12.1}$$

where $\vec{R}_{\ell mn} = \ell\vec{a} + m\vec{b} + n\vec{c} + \vec{\rho}_j$, and $\vec{a}$, $\vec{b}$, and $\vec{c}$ are lattice vectors. The charge at site j is eZ_j, and $\vec{r}$ is the position of an electron on the ion being discussed. The multipolar expansion of equation (12.1) is

$$H_3 = \sum_{\ell mn} \sum_j \frac{-e^2 Z_j r^k}{[R_{\ell m}(j)]^{k+1}} C_{kq}(\hat{r}) C_{kq}^*[\hat{R}_{\ell mn}(j)] \quad . \tag{12.2}$$

The multipolar crystal-field components A_{kq} are

$$A_{kq} = -e^2 \sum_{\ell mn} \sum_j \frac{Z_j C_{kq}[\hat{R}_{\ell mn}(j)]}{[R_{\ell mn}(j)]^{k+1}} \quad . \tag{12.3}$$

Thus the point-charge Hamiltonian is

$$H_3 = \sum_{kq} A_{kq}^* \sum_{i=1}^N r_i^k C_{kq}(\hat{r}_i) \quad , \tag{12.4}$$

where we have summed over all the N electrons in the $4\ell^N$ configuration. If all the lengths are measured in angstroms (10^{-8} cm), then

$$A_{kq} = - \frac{\alpha_o}{2\pi} \times 10^8 \sum_{\ell mn} \sum_j \frac{z_j C_{kq}[\hat{R}_{\ell mn}(j)]}{[R_{\ell mn}(j)]^{k+1}} \quad , \qquad (12.5)$$

where the fine-structure constant, α_o, is $e^2/\hbar c$ (thus, $\alpha_o/2\pi \times 10^8 = 116,140$), and the units of A_{kq} are $cm^{-1}/\AA^k$. If $\langle r^k \rangle$ is in angstroms, then $A_{kq}\langle r^k \rangle$ is in cm^{-1}.

The sum in equation (12.5) always converges--even for the lowest k value (k = 0)--if taken in the order indicated. That is, the sum over j is performed with ℓ, m, and n fixed. The unit cell is neutral, that is,

$$\sum_j z_j = 0 \quad . \qquad (12.6)$$

In many (but not all) space groups, it is possible to choose an origin for the lattice coordinates such that the dipole moment of the unit cell vanishes; that is,

$$\sum_j \vec{\rho}_j z_j = 0 \quad , \qquad (12.7)$$

where $\vec{\rho}_j$ is the position of the jth charge in the unit cell. The result in equation (12.7) can be anticipated by observing the point symmetry of the ions in a specific solid. If the ions occupy C_1, C_2, C_s, C_{2v}, C_4, C_{4v}, C_3, C_{3v}, C_6, or C_{6v} point symmetry (Schoenflies notation), then it is impossible to satisfy equation (12.7) with these sites as the origin in a unit cell. If it is possible to satisfy equation (12.7), then the sum given in equation (12.5) converges very rapidly. This can be shown from the expansion

$$\frac{C_{kq}(\widehat{R-x})}{|\vec{R}-\vec{x}|^{k+1}} = \sum_{a\alpha} \binom{2a+2k}{2a}^{1/2} \langle a(\alpha)k(q)|a+k(\alpha+q)\rangle \, x^a C_{a\alpha}(\hat{x}) \frac{C^*_{a+k,\alpha+q}(\hat{R})}{R^{a+k+1}} \qquad (12.8)$$

(Carlson and Rushbrooke, 1950). With $\vec{\rho}_j = \vec{x}$ and $\vec{R}_{\ell mn}(0) = \vec{R}$ (where $\vec{R}_{\ell mn}(j) = \vec{R}_{\ell mn}(0) + \vec{\rho}_j$), for the sum in equation (12.5) we have

$$\sum_{\ell mn} \sum_j \frac{Z_j C_{kq}[\hat{R}_{\ell mn}(j)]}{[R_{\ell mn}(j)]^{k+1}} = \sum_{\ell mn} \sum_{a\alpha} \left(\frac{2a+2k}{2a}\right)^{1/2} \langle a(\alpha)k(q)|a+k(\alpha+q)\rangle$$

$$\tag{12.9}$$

$$\times \sum_j Z_j \rho_j^a C_{a\alpha}(\hat{\rho}_j) \frac{C^*_{a+k,\alpha+q}[\vec{R}_{\ell mn}(0)]}{[R_{\ell mn}(0)]^{a+k+1}} \quad .$$

Now if equation (12.7) is satisfied, then

$$\sum_j Z_j \rho_j C_{1\alpha}(\hat{\rho}_j) = 0 \quad . \tag{12.10}$$

Thus, the sum in equation (12.9) is for $a > 1$; we see that even for the lowest term, $k = 0$, for large ℓ, m, and n the individual terms decrease as $1/R^2_{\ell mn}(0)$. While the expansions in equations (12.8) and (12.9) are good for demonstrating the rate of convergence, the computation of A_{kq} by equation (12.5) is more practical. However, in equation (12.5), the sum over j should be done for each cell first, with fixed values of ℓ, m, and n. In programming language, this is expressed by stating that the j loop is the innermost of the nested ℓ, m, n, and j loops. In some lattices, the condition in equation (12.7) may place some of the point charges on the cell faces. In these cases it is a simple matter to balance these charges by an adjustment to fractions of equal charges on opposite faces.

The convention we use for our lattice sums is that given in the International Tables for Crystallography (1952); table 12.1 is reproduced from volume I (the other two volumes give data strictly for x-ray crystallographers). The data used in the lattice sums are generally those reported in Acta Crystallographica, section B (now predominately section C). Care should be taken to make certain the correct setting is used.

Typical data used in the calculation of the lattice sums are given in table 12.2 (LiYF$_4$, calcium tungstate space group 88, scheelite structure). All the data given in table 12.2 are given in the International Tables, except that the x, y, and z coordinates are determined by x-ray diffraction. The lattice constants a, b, and c are also determined by x-ray diffraction and, as customary, the true positions of the ions are xa, yb, and zc (these relations hold for all the ions in a unit cell). The polarizability (from Kittel, 1956, p 165) of each ion is given at the bottom of table 12.2. For this particular solid, only the fluorine ions can have dipole moments that contribute to the crystal field (we discuss the role of the dipole moments in sect. 14). Not all the data for space group 88 are contained in table 12.2 because the

TABLE 12.1.   CRYSTALLOGRAPHIC AXIAL AND ANGULAR RELATIONSHIPS
AND CHARACTERISTIC SYMMETRY OF CRYSTAL SYSTEMS

Space group	System	Axial and angular relationships	X-ray data needed for unit cell
1, 2	Triclinic	$a \neq b \neq c$ $\alpha \neq \beta \neq \gamma \neq 90°$	$a, b, c, \alpha, \beta, \gamma$
3 to 5	Monoclinic	First setting $a \neq b \neq c$ $\alpha = \beta = 90° \neq \gamma$	a, b, c, γ
		Second setting $a \neq b \neq c$ $\alpha = \gamma = 90° \neq \beta$	a, b, c, β
16 to 74	Orthorhombic	$a \neq b \neq c$ $\alpha = \beta = \gamma \neq 90°$	a, b, c
75 to 142	Tetragonal	$a = b \neq c$ $\alpha = \beta = \gamma = 90°$	a, c
143 to 167	Trigonal (may be taken as subdivision hexagonal)	(Rhombohedral axes) $a = b = c$ $\alpha = \beta = \gamma < 120° \neq 90°$ $a = b \neq c$ $\alpha = \beta = 90°$ $\gamma = 120°$	a, α
168 to 194	Hexagonal	$a = b \neq c$ $\alpha = \beta = 90°$ $\gamma = 120°$	a, c
195 to 230	Cubic	$a = b = c$ $\alpha = \beta = \gamma = 90°$	a

Source: International Tables, 1952, Vol. I, p 11, table 2.3.1.

TABLE 12.2.   CRYSTALLOGRAPHIC DATA FOR $LiYF_4$
(SCHEELITE, $CaWO_4$), TETRAGONAL SPACE
GROUP 88 (FIRST SETTING), Z = 4

Ion	Position	Symmetry	x	y	z
Y	4b	S_4	0	0	1/2
Li	4a	S_4	0	0	0
F	16f	C_1	0.2820	0.1642	0.0815

Note: $a = 5.1668$ Å, $b = a$, $c = 10.733$ Å, $\alpha = 90°$, $\beta = 90°$, $\gamma = 90°$, $\alpha_Y = 0.55$ Å^3, $\alpha_{Li} = 0.05$ Å^3, $\alpha_F = 1.04$ Å^3 (reduced to 0.104 in the lattice sum), $Z_Y = +3$, $Z_{Li} = +1$, $Z_F = -1$.

equivalent positions given in the International Tables are generated inside the program. The centering position in the cell can be taken as either the Y or Li site, since these positions have S_4 symmetry, and their lowest crystal-field component is A_{20}. Equation (12.10) is therefore automatically satisfied. The resulting lattice sum for the Y site in $LiYF_4$ for the parameters in table 12.2 is given in table 12.3. The sum covers all the complete cells in a sphere of 30-A radius and should be an accurate result. Also included in table 12.3 are the results for the dipole contributions due to the presence of dipoles at the fluorine sites (see sect. 14).

As a second example, we choose a very low symmetry solid, YCl_3, which is characterized by monoclinic space group 12. As can be seen in table 12.4, all the ions are in very low point-symmetry positions, and each position can have a dipole moment (another way of saying this is that each position has a onefold field, A_{1m}). We then have to consult the International Tables for a higher symmetry position in order to satisfy equation (12.7), which in this case is the site 4e with C_i symmetry. The C_i point group has only the inversion operation, and all the odd-n A_{nm} vanish in this symmetry. Thus if the position 4e is used, equation (12.7) will automatically be satisfied. The lattice sum for YCl_3 was also run over a lattice 30 × 30 × 30, and only the even-n A_{nm} are given in table 12.5. The dipole contributions were also calculated; these calculations were more complicated in this solid because of the three types of sites (Y, Cl_1, Cl_2); all have a dipole moment. For many of the A_{nm}, the dipole contributions are much larger than the monopole terms. This frequently happens when the handbook values for the dipole polarizabilities are used. We have had more believable results when we reduce the polarizability to one tenth of the handbook value.

The lattice sums given in tables 12.3 and 12.5 are incomplete in that the results are not in a usable form for many of our computer programs. Before we can use these results, the A_{nm} should be rotated using the standard

TABLE 12.3. LATTICE SUMS FOR Y SITE AT (0, 0, 1/2) FOR $LiYF_4$ WITH Z_Y = 3, Z_{Li} = -1, Z_F = -1, α_F = 0.104 A^3

Lattice sum	Monopole A_{nm}		Dipole A_{nm}		Monopole and dipole	
	Real	Imaginary	Real	Imaginary	Real	Imaginary
A_{20}	1074	0	340	0	1414	0
A_{32}	373	859	-358	74.0	15	933
A_{40}	-1957	0	-98.1	0	-2055	0
A_{44}	-2469	-2362	-3.83	-80.3	-2473	-2442
A_{52}	1050	-2456	1.28	-74.7	1051	-2531
A_{60}	-17.2	0	7.96	0	-9.24	0
A_{64}	-615	-420	-29.03	-9.37	-644	-429
A_{72}	-15.7	0.90	1.55	-9.94	-14.2	-9.04
A_{76}	250	-63.8	17.8	7.96	268	-55.9

TABLE 12.4. CRYSTALLOGRAPHIC DATA
FOR YCl$_3$, MONOCLINIC SPACE GROUP 12
(C2/m) (SECOND SETTING), Z = 4

Ion	Position	Symmetry	x	y	z
Y	4g	C_2	0	0.166	0
Cl$_1$	4i	C_s	0.211	0	0.247
Cl$_2$	8j	C_1	0.229	0.179	0.760
--	4e	C_i	1/4	1/4	0

*Note: a = 6.92 Å, b = 11.94 Å, c = 6.44 Å,
α = 90°, β = 11.0°, γ = 90°. Charges: q_Y = 3,
q_{Cl} = -1.
Polarizability: a_Y = 0.55 Å^3, a_{Cl} = 3.66 Å^3.*

TABLE 12.5. LATTICE SUMS FOR Y SITE AT
(0, 0.166, 0) FOR YCl$_3$, EVEN-n A_{nm}
ONLY, ALL A_{nm} REAL

Lattice sum	Monopole	Dipole	Total
A_{20}	1738	3227	4965
A_{21}	-913	2916	2003
A_{22}	245	2574	2819
A_{40}	-73.9	246	172
A_{41}	85.8	-398	-312
A_{42}	-41.3	47.7	6.4
A_{43}	10.4	-791	-781
A_{44}	-3.64	516	512
A_{60}	-0.06	-80.2	-80.3
A_{61}	-3.76	21.3	17.5
A_{62}	3.35	-27.4	-24.0
A_{63}	-0.58	60.5	59.9
A_{64}	2.73	39.2	41.9
A_{65}	5.07	13.7	18.8
A_{66}	8.14	-65.9	-57.8

Euler angle-rotation matrix, so that the lattice sums, A'_{nm}, rotated from A_{nm} by the angles α, β, and γ, are

$$A'_{nm} = \sum_m D^n_{m'm}(\alpha,\beta,\gamma) A_{nm'} \quad .$$

Explicit forms for the $D^n_{m'm}(\alpha,\beta,\gamma)$ are given in Rose (1957, ch. IV).

12.2 Bibliography and References

Bethe, H. A. (1929), Termaufspaltung in Kristallen (translation: Splitting of Terms in Crystals, by Consultants Bureau, Inc., New York, NY), Ann. Physik (Leipzig) 3, 133.

Bethe, H. (1930), Zur Theorie des Zeemaneffektes an den Salzen der seltenen Erden, Z. Physik 60, 218.

Carlson, B. C., and G. S. Rushbrooke (1950), On the Expansion of a Coulomb Potential in Spherical Harmonics, Proc. Camb. Phil. Soc. 46, 626.

Faucher, M., and P. Caro (1977), A Quickly Converging Method for Computing Electrostatic Crystal Field Parameters, J. Chem. Phys. 66, 1273.

Faucher, M., and D. Garcia (1983a), Crystal Field Effects in Rare-Earth-Doped Oxyhalides: Ab-Initio Calculations Including Effects of Dipolar and Quadrupolar Moments, Solid State Chemistry, Proceedings of the Second European Conference (7-9 June 1982, Veldhoven, Netherlands), Elsevier, Amsterdam, Netherlands, p 547.

Faucher, M., and D. Garcia (1983b), Crystal Field Effects on 4f Electrons: Theories and Reality, J. Less-Common Metals 93, 31.

Faucher, M., and D. Garcia (1982), Electrostatic Crystal-Field Contributions in Rare-Earth Compounds with Consistent Multipolar Effects: I--Contribution to K-Even Parameters, Phys. Rev. B 26, 5451.

Garcia, D. (1983), Simulation ab-initio des parametres du champ des ligandes, thesis, L'Ecole Centrale des Arts et Manufactures.

Garcia, D., and M. Faucher (1984), Crystal-Field Parameters in Rare-Earth Compounds: Extended Charge Contributions, Phys. Rev. A 30, 1703.

Garcia, D., M. Faucher, and O. L. Malta (1983), Electrostatic Crystal-Field Contributions in Rare-Earth Compounds with Consistent Multipolar Effects: II.--Contribution to K-Odd Parameters (Transition Probabilities), Phys. Rev. B 27, 7386.

Hutchings, M. T., and D. K. Ray (1963), Investigations into the origin of Crystalline Electronic Field Effects on Rare-Earth Ions: I--Contributions from Neighbouring Induced Moments, Proc. Phys. Soc. 81, 663.

International Tables for X-Ray Crystallography (1952), Vol. I.

Karayianis, N., and C. A. Morrison (1973), Rare-Earth Ion-Host Lattice Interactions: 1.--Point Charge Lattice Sum in Scheelites, Harry Diamond Laboratories, HDL-TR-1648.

Kittel, C. (1956), Introduction to Solid State Physics (2nd ed.), Wiley, New York, NY.

Leavitt, R. P., and C. A. Morrison (1980, 15 July), Crystal Field Analysis of Triply Ionized Rare Earth Ions in Lanthanum Trifluoride: II--Intensity Calculations, J. Chem. Phys. 73, 749.

Morrison, C. A. (1976), Dipolar Contributions to the Crystal Fields in Ionic Solids, Solid State Commun. 18, 153.

Morrison, C. A., and R. P. Leavitt (1982), Spectroscopic Properties of Triply Ionized Lanthanides in Transparent Host Materials, in Volume 5, Handbook of the Physics and Chemistry of Rare Earths, ed. by K. A. Gschneidner, Jr., and L. Eyring, North-Holland Publishers, New York, NY.

Morrison, C. A., and R. P. Leavitt (1979, 15 September), Crystal Field Analysis of Triply Ionized Rare Earth Ions in Lanthanum Trifluoride, J. Chem. Phys. 71, 2366.

Morrison, C. A., G. F. de Sa, and R. P. Leavitt (1982), Self-Induced Multipole Contribution to the Single-Electron Crystal Field, J. Chem. Phys. 76, 3899.

Rose, M. E. (1957), Elementary Theory of Angular Momentum, Wiley, New York, NY.

13. POINT-CHARGE MODEL DEVELOPED AT HDL

In the classical point-charge model, the crystal-field parameters, B_{nm}, for the crystal-field interaction of the form

$$H_3 = \sum_{nm} B^*_{nm} \sum_i C_{nm}(i) \tag{13.1}$$

were calculated as

$$B_{nm} = \langle r^n \rangle A_{nm} \quad , \tag{13.2}$$

where the $\langle r^n \rangle$ are the expectation values of r^n of the rare-earth ion, and the A_{nm} are the multipole components of the energy at the site occupied by the rare-earth ion. In the earlier models, the radial integrals used in the evaluation of r^n were taken from Hartree-Fock calculations (Freeman and Watson, 1962), and the A_{nm} were calculated using the point charges at the valence values for the constituent ions. These calculations generally gave the twofold field 10 times too large, the fourfold fields approximately in good agreement, and the sixfold fields 10 times too small.

13.1 Screening and Wave Function Spread

Several errors in the classical theory were immediately obvious. If the radial wave functions (Hartree-Fock) for the free ion were correct, then these wave functions should give the correct values for the Slater integrals F^2, F^4, and F^6. They did not for Pr^{3+}. A simple procedure was then applied. The radial wave functions were assumed to be of the form

$$\phi(r) = CR_{HF}(\tau r) \tag{13.3}$$

where τ is a parameter, C is a normalization factor, and $R_{HF}(r)$ are the Hartree-Fock radial wave functions. With the radial function given by equation (13.3), the Slater integrals become

$$F^k = \tau F^k_{HF} \quad , \tag{13.4}$$

and it was found that a value of τ of approximately 0.75 was needed to fit the F^k that are found by fitting the experimental spectra of Pr^{3+}. Thus, the Hartree-Fock radial wave functions had their maxima too near the origin and needed to be spread out even in the free ion, and perhaps more spreading was necessary for an ion in a solid.

From radial wave functions given in equation (13.3), it is not difficult to show that

$$\langle f(r) \rangle = \langle \phi | f(r) | \phi \rangle / \langle \phi | \phi \rangle$$

$$(13.5)$$

$$= \langle f(r/\tau) \rangle_{HF} \quad ,$$

so that any quantity that has been calculated using Hartree-Fock functions is immediately obtained, particularly

$$\langle r^k \rangle = \langle r^k \rangle_{HF} / \tau^k \quad . \qquad (13.6)$$

A second error of the classical method was the omission of the Sternheimer shielding factors (Sternheimer, 1951, 1966; Sternheimer et al, 1968). In 1952 Sternheimer showed that, in a multipolar expansion of the energy of a point charge embedded in a solid, the r^n should be replaced by $r^n(1 - \sigma_n)$, where the σ_n are the shielding factors. He further showed that these factors were independent of azimuthal angle; that is, if the angular variation in the multipolar expansion was given by Y_{nm}, the σ_n were independent of m. The values of σ_n have been calculated for Pr^{3+} and Tm^{3+} and are

$$\sigma_2 = 0.666 \quad , \quad \sigma_4 = 0.09 \quad , \quad \sigma_6 = 0.04 \quad \text{for } Pr^{3+} \quad ,$$

$$(13.7)$$

$$\sigma_2 = 0.545 \quad , \quad \sigma_4 = 0.09 \quad , \quad \sigma_6 = 0.04 \quad \text{for } Tm^{3+} \quad ,$$

where the replacement is

$$r^n \rightarrow r^n(1 - \sigma_n) \quad . \qquad (13.8)$$

More recent calculations of the shielding factors have been done (Sengupta and Artman, 1970, and perhaps others) which we shall need if further refinements of the theory are undertaken.

13.2 Effective Charge and Position

The crystal-field components, A_{nm}, are a function of the position of the ions in a solid; in solids such as $CaWO_4$ the $(WO_4)^{-2}$ complex is known to be covalent. That is, the charges on the tungsten and the oxygen ions are not necessarily at their valence values. If we let the charge on the tungsten ion be q_W, then we require that

$$q_W + 4q_O = -2 \qquad (13.9)$$

with the charge on the oxygen being q_O. The result given in equation (13.9) then assumes that the Ca^{2+} site is purely ionic with charge 2. This assumption is consistent with many of the experimental results on compounds such as $CaWO_4$ or YVO_4. We introduce a second parameter, the effective position of the oxygen ion relative to the tungsten site that would reproduce the effective dipole moment seen from the Ca^{2+} site. This parameter, η, is introduced by

$$R_{O-W}(\text{effective}) = \eta R_{O-W}(\text{measured}) \quad , \qquad (13.10)$$

where R_{O-W} is the distance from the oxygen nucleus to the tungsten nucleus. Thus there are two parameters in the A_{nm}: q_O, the effective charge (q_W is eliminated by eq (13.9)), and η, the effective distance of the oxygen site from the tungsten site. The calculated crystal-field parameters B_{nm} then are

$$B_{nm}(\tau; q_O, \eta) = \langle r^n \rangle_{HF}(1 - \sigma_n)A_{nm}(q_O, \eta)/\tau^n \quad , \qquad (13.11)$$

with the three parameters τ, q_O, and η.

The experimental data that were taken at HDL on the rare-earth ions in $CaWO_4$ were analyzed using the effective spin-orbit Hamiltonian (Karayianis, 1970), and a set of phenomenological B_{nm} was obtained. These, given in table 13.1, are the B_{nm} that the theory has to fit.

The fitting was done by minimizing the square quantity given by

$$Q = \sum_{nm} \left[B_{nm}(\tau; q_O, \eta) - B_{nm} \right]^2 \quad , \qquad (13.12)$$

where $B_{nm}(\tau; q_O, \eta)$ is given by equation (13.11), and B_{nm} is from table 13.1, for each ion. The minimization was done with respect to τ, q_O, and η for each ion. Since the q_O and η are assumed to be ion independent and τ is assumed to be host independent, the average q_O and η were taken and fixed. The process was then repeated with minimization with respect to τ only. This process yielded the following:

$$q_O = -1.09 \quad , \qquad \eta = 0.977 \quad . \qquad (13.13)$$

TABLE 13.1. PHENOMENOLOGICAL B_{nm} FOR SIX RARE-EARTH IONS IN $CaWO_4$ (all in cm^{-1})

Ion	B_{20}	B_{40}	B_{44}	B_{60}	B_{64}	
					Re	Im
Nd	549	-942	1005	-17	947	1
Tb	468	-825	872	-290	595	160
Dy	428	-825	978	-7	448	2.5
Ho	436	-664	779	-33	558	196
Er	404	-685	728	12	452	164
Tm	417	-688	754	17	506	359

The τ values were well fitted by

$$\tau = 0.767 - 0.00896N \quad , \qquad (13.14)$$

where N is the number of f electrons in the configuration $4f^N$. The values of σ_n used in the above were not varied in the minimizing process but were interpolated from the values given in equation (13.7); that is,

$$\sigma_2 = 0.6902 - 0.0121N \quad ,$$

$$\sigma_4 = 0.09 \quad \text{(all N)} \quad , \qquad (13.15)$$

$$\sigma_6 = -0.04 \quad \text{(all N)} \quad .$$

The predicted values of the $B_{nm}(\tau; q_0, \eta)$ for the entire rare-earth series are given in table 13.2. The results given in table 13.2 when compared to table 13.1 show that the difference between the derived $B_{nm}(\tau; q_0, \eta)$ and the phenomenological B_{nm} is greater for the low-N ions in the $4f^N$ configuration. This may be a defect in the theory, but not enough data on the low-N ions are available for analysis. One of the significant results of the analysis was that it led to the reanalysis of the spectrum of Tb:CaWO$_4$ with a different interpretation of the experimental data (Leavitt et al, 1974).

More recent work on CaWO$_4$ (Morrison et al, 1977) obtains the following values:

$$\sigma_2 = 0.6846 - 0.00854N,$$

$$\sigma_4 = 0.02356 + 0.00182N,$$

$$\tau = 0.75(1.0387 - 0.0129N),$$

$$\sigma_6 = -0.04238 + 0.00014N,$$

$$q_0 = -1.150, \quad \text{and}$$

$$\eta = 0.962.$$

The σ_n values are interpolated from the calculations of Erdos and Kang (1972) for Pr^{3+} and Tm^{3+}. The factors in equation (13.11) were combined so that

$$\rho_n = \langle r^n \rangle (1 - \sigma_n)/\tau^n \quad , \qquad (13.16)$$

and the ρ_n along with the τ are given in table 13.3. Thus we have

$$B_{nm}(\tau; q_0, \eta) = \rho_n A_{nm}(q_0, \eta) \quad . \qquad (13.17)$$

At present we use the results given in table 13.3 to calculate crystal-field parameters given by equation (13.17) and use these parameters as starting values to best fit experimental data. We generally use $A_{nm}(q,\eta)$ with $\eta = 1$ in the process (q here is the effective charge on the ligands, not necessarily oxygen). After obtaining the best-fit B_{nm}, we return to the calculation of $A_{nm}(q)$ and vary q to obtain the best fit by minimizing the quantity

$$Q = \sum_{nm} \left[B_{nm} - \rho_n A_{nm}(q) \right]^2 . \tag{13.18}$$

Following this, we obtain the $A_{nm}(q)$ for odd n and use them to calculate the intensities using the Judd-Ofelt theory.

TABLE 13.2. DERIVED CRYSTAL-FIELD PARAMETERS, $B_{nm}(\tau;q_0,n)$ FOR $4f^N$ CONFIGURATION OF TRIPLY IONIZED RARE-EARTH IONS (all in cm^{-1})

N	Ion	B_{20}	B_{40}	B_{44}	B_{60}	B_{64}	
						Re	Im
1	Ce	441	-1429	1462	16	1251	52
2	Pr	424	-1224	1253	13	996	42
3	Nd	408	-1059	1083	11	805	34
4	Pm	411	-1017	1041	10	764	32
5	Sm	408	-938	960	9	676	28
6	Eu	408	-887	908	8	626	26
7	Gd	406	-824	843	7	559	23
8	Tb	424	-856	876	8	591	25
9	Dy	428	-831	851	7	563	24
10	Ho	417	-756	774	6	488	20
11	Er	415	-707	724	6	439	18
12	Tm	435	-729	746	6	454	19
13	Yb	434	-701	717	6	429	18

TABLE 13.3. VALUES FOR τ, $\langle r^n \rangle_{HF}$, σ_n, AND ρ_n FOR $4f^N$ CONFIGURATION OF TRIPLY IONIZED RARE-EARTH IONS[a]

Ion	N	τ	$\langle r^2 \rangle_{HF}$	$\langle r^4 \rangle_{HF}$	$\langle r^6 \rangle_{HF}$	σ_2	σ_4	σ_6	ρ_2	ρ_4	ρ_6
Ce	1	0.7693	0.3360	0.2709	0.4659	0.6757	0.0254	-0.0422	0.1841	0.7536	2.3417
Pr	2	0.7597	0.3041	0.2213	0.3459	0.6667	0.0272	-0.0421	0.1756	0.6464	1.8754
Nd	3	0.7500	0.2803	0.1882	0.2715	0.6578	0.0290	-0.0420	0.1706	0.5776	1.5897
Pm	4	0.7403	0.2621	0.1655	0.2247	0.6488	0.0308	-0.0418	0.1679	0.5339	1.4213
Sm	5	0.7306	0.2472	0.1488	0.1929	0.6298	0.0327	-0.0417	0.1668	0.5049	1.3210
Eu	6	0.7210	0.2347	0.1353	0.1686	0.6309	0.0345	-0.0415	0.1666	0.4836	1.2503
Gd	7	0.7113	0.2232	0.1237	0.1477	0.6220	0.0363	-0.0414	0.1668	0.4656	1.1873
Tb	8	0.7016	0.2129	0.1131	0.1287	0.6130	0.0381	-0.0413	0.1673	0.4990	1.1232
Dy	9	0.6919	0.2033	0.1037	0.1119	0.6041	0.0399	-0.0411	0.1681	0.4341	1.0614
Ho	10	0.6823	0.1945	0.0945	0.0981	0.5951	0.0418	-0.0410	0.1692	0.4217	1.0119
Er	11	0.6726	0.1865	0.0883	0.0874	0.5861	0.0436	-0.0408	0.1706	0.4126	0.9826
Tm	12	0.6629	0.1790	0.0820	0.0787	0.5772	0.0454	-0.0407	0.1722	0.4053	0.9649
Yb	13	0.6532	0.1717	0.0753	0.0681	0.5683	0.0472	-0.0406	0.1737	0.3938	0.9120

[a] The units of $\langle r^n \rangle_{HF}$ and p_n are $Å^n$.

At present we have not included the dipole contribution to the $A_{nm}(q)$ but intend to do so as soon as possible. The old η in the three-parameter theory will be replaced by α, the polarizability of the constituent ions in low-symmetry sites. We believe that this procedure (including finding new ρ_n values) will give much better results than obtained by the older theory. In our projected reanalysis we will have the good phenomenological B_{nm} for $R^{3+}:LaF_3$, $R^{3+}:LaCl_3$, and $R^{3+}:LiYF_4$ (these are reported by Morrison and Leavitt, 1982), and will soon have $R^{3+}:Y_2O_3$, in addition to the B_{nm} for $R^{3+}:CaWO_4$ used in the older theory. These data should be sufficient to form a very stringent test of a newer three-parameter theory.

For the nd^N ions (X^q for q = +2, +3, and +4,) we have taken the values of $F^{(k)}$ obtained by fitting the free-ion data (Uylings et al, 1984) along with the Hartree-Fock values of $F^{(k)}$ (Fraga et al, 1976) and obtained values of τ using equation (13.4). Using these values of τ, we obtained the estimated values of $\langle r^k \rangle$ from equation (13.6). The resulting $\langle r^k \rangle$ are given in table 13.4.

The results given in table 13.4 along with reported values of A_{kq} (Morrison and Schmalbach, 1985) can be used in equation (13.2) to obtain crystal-field parameters, B_{kq}, which can be used as starting values in fitting the experimental data. A similar process can be performed for the X^{+q} (q = 2, 3, and 4) for the $4d^N$ and $5d^N$ series. However, because of the lack of free-ion parameters $F^{(k)}$ for these ions, we shall have to interpolate from the $3d^N$ series.

TABLE 13.4. ESTIMATED VALUES OF $\langle r^k \rangle$ (A^k) DIVALENT, TRIVALENT, AND QUADRIVALENT IONS WITH $3d^N$ ELECTRONIC CONFIGURATION

nd^N	X^{2+}	$\langle r^2 \rangle^a$	$\langle r^4 \rangle^a$	X^{3+}	$\langle r^2 \rangle^b$	$\langle r^4 \rangle^b$	X^{4+}	$\langle r^2 \rangle^c$	$\langle r^4 \rangle^c$
$3d^1$	Sc	1.372	4.053	Ti	0.7958	1.281	V	0.6217	1.298
$3d^2$	Ti	1.073	2.505	V	0.6689	0.9145	Cr	0.5172	0.8177
$3d^3$	V	0.8822	1.718	Cr	0.5776	0.6911	Mn	0.4911	0.7761
$3d^4$	Cr	0.7423	1.234	Mn	0.5054	0.5363	Fe	0.3958	0.4955
$3d^5$	Mn	0.6293	0.8973	Fe	0.4436	0.4177	Co	0.3648	0.4081
$3d^6$	Fe	0.5576	0.7236	Co	0.4020	0.3506	Ni	0.3304	0.3282
$3d^7$	Co	0.4917	0.5738	Ni	0.3627	0.2903	Cu	0.2982	0.2600
$3d^8$	Ni	0.4353	0.4577	Cu	0.3280	0.2413	Zn	0.2708	0.1995
$3d^9$	Cu	0.3871	0.3678	Zn	0.2977	0.2016	Ga	0.2291	0.1296

$^a \langle r^k \rangle_{HF}/\tau^k$ calculated using $\tau = 0.76878 + 0.11128N$.
$^b \langle r^k \rangle_{HF}/\tau^k$ calculated using $\tau = 0.811184 + 0.0073953N$.
$^c \langle r^k \rangle_{HF}/\tau^k$ calculated using $\tau = 0.833540 + 0.0056609N$.

13.3 Annotated Bibliography and References

E. A. Brown, J. Nemarich, N. Karayianis, and C. A. Morrison (1970, November), Evidence for Yb^{3+} 4f Radial Wavefunction Expansion in Scheelites, Phys. Lett. 33A, 375. In this article the wavefunction expansion was assumed to be due to the solid; however, the analysis of the Pr^{3+} spectrum showed that the expansion parameter is needed to bring the Hartree $F^{(k)}$ into agreement with the observed $F^{(k)}$. To a certain extent this is true of the $3d^N$ series, as can be seen by the values of τ given in table 13.4.

Erdos, P., and J. H. Kang (1972), Electric Shielding of Pr^{3+} and Tm^{3+} Ions in Crystals, Phys. Rev. B6, 3383.

Fraga, S., K. M. S. Saxena, and J. Karwowski (1976), Handbook of Atomic Data, Elsevier, New York, NY.

Freeman, A. J., and R. E. Watson (1962), Theoretical Investigation of Some Magnetic and Spectroscopic Properties of Rare-Earth Ions, Phys. Rev. 127, 2058.

Karayianis, N. (1971), Theoretical Energy Levels and g Values for the ^{4}I Terms of Nd^{3+} and Er^{3+} in $LiYF_4$, J. Phys. Chem. Solids 32, 2385.

Karayianis, N. (1970, 15 September), Effective Spin-Orbit Hamiltonian, J. Chem. Phys. 53, 2460.

Karayianis, N., and R. T. Farrar (1970, November), Spin-Orbit and Crystal Field Parameters for the Ground Term of Nd^{3+} in $CaWO_4$, J. Chem. Phys. 53, 3436.

Karayianis, N., and C. A. Morrison (1975, January), Rare Earth Ion-Host Crystal Interactions: 2.--Local Distortion and Other Effects in Reconciling Lattice Sums and Phenomenological B_{km}, Harry Diamond Laboratories, HDL-TR-1682.

Karayianis, N., and C. A. Morrison (1973, October), Rare Earth Ion-Host Lattice Interactions. 1.--Point Charge Lattice Sum in Scheelites, Harry Diamond Laboratories, HDL-TR-1648.

Leavitt, R. P., C. A. Morrison, and D. E. Wortman (1975, June), Rare Earth Ion-Host Crystal Interactions: 3.--Three-Parameter Theory of Crystal Fields, Harry Diamond Laboratories, HDL-TR-1673.

Leavitt, R. P., C. A. Morrison, and D. E. Wortman (1974), Description of the Crystal Field for Tb^{3+} in $CaWO_4$, J. Chem. Phys. 61, 1250.

Morrison, C. A., N. Karayianis, and D. E. Wortman (1977, June), Rare Earth Ion-Host Lattice Interactions: 4.--Predicting Spectra and Intensities of Lanthanides in Crystals, Harry Diamond Laboratories, HDL-TR-1816.

Morrison, C. A., and R. P. Leavitt (1982), Spectroscopic Properties of Triply Ionized Lathanides in Transparent Host Materials, in Volume 5, Handbook of the Physics and Chemistry of Rare Earths, ed. by K. A. Gschneidner, Jr., and L. Eyring, North-Holland Publishers, New York, NY.

Morrison, C. A., R. P. Leavitt, and A. Hansen (1985, October), Host Materials for Transition-Metal Ions with the nd^N Electronic Configuration, Harry Diamond Laboratories, HDL-DS-85-1.

Morrison, C. A., D. R. Mason, and C. Kikuchi (1967), Modified Slater Integrals for an Ion in a Solid, Phys. Lett. 24A, 607.

Morrison, C. A., and R. G. Schmalbach (1985, July), Approximate Values of $\langle r^k \rangle$ for the Divalent, Trivalent, and Quadrivalent Ions with the $3d^N$ Electronic Configuration, Harry Diamond Laboratories, HDL-TL-85-3.

Morrison, C. A., and D. E. Wortman (1971, November), Free Ion Energy Levels of Triply Ionized Thulium Including the Spin-Spin, Orbit-Orbit, and Spin-Other-Orbit Interactions, Harry Diamond Laboratories, HDL-TR-1563.

Sengupta, D., and J. O. Artman (1970), Crystal-Field Shielding Parameters for Nd^{3+} and Np^{4+}, Phys. Rev. B1, 2986.

Stephens, R. R., and D. E. Wortman (1967, November), Comparison of the Ground Term, Energy Levels, and Crystal Field Parameters of Terbium in Scheelite Crystals, Harry Diamond Laboratories, HDL-TR-1367.

Sternheimer, R. M. (1966), Shielding and Antishielding for Various Ions and Atomic Systems, Phys. Rev. 146, 140.

Sternheimer, R. M. (1951), Nuclear Quadrupole Moments, Phys. Rev. 84, 244.

Sternheimer, R. M., M. Blume, and R. F. Peierls (1968), Shielding of Crystal Fields at Rare-Earth Ions, Phys. Rev. 173, 376.

Uylings, P. H. M., A. J. J. Raassen, and J. F. Wyart (1984), Energies of N Equivalent Electrons Expressed in Terms of Two-Electron Energies and Independent Three-Electron Parameters: A New Complete Set of Orthogonal Operators: II.--Application of $3d^N$ Configurations, J. Phys. B17, 4103.

Wortman, D. E. (1972), Ground Term Energy States for Nd^{3+} in $LiYF_4$, J. Phys. Chem. Solids 33, 311.

Wortman, D. E. (1971a), Optical Spectrum of Triply Ionized Erbium in Calcium Tungstate, J. Chem. Phys. 54, 314.

Wortman, D. E. (1971b), Ground Term of Nd^{3+} in $LiYF_4$, Bull. Am. Phys. Soc. 16, 594.

Wortman, D. E. (1970a), Ground Term Energy Levels and Possible Effect on Laser Action for Er^{3+} in $CaWO_4$, J. Opt. Soc. Am. 60, 1143.

Wortman, D. E. (1970b, June), Optical Spectrum of Triply Ionized Erbium in Calcium Tungstate, Harry Diamond Laboratories, HDL-TR-1510.

Wortman, D. E. (1968a), Analysis of the Ground Term of Tb^{3+} in $CaWO_4$, Phys. Rev. 175, 488.

Wortman, D. E. (1968b), Absorption and Fluorescence Spectra and Crystal Field Parameters of Tb^{3+} in $CaWO_4$, Bull. Am. Phys. Soc. 13, 686.

Wortman, D. E. (1968c, February), Absorption and Fluorescence Spectra and Crystal Field Parameters of Triply Ionized Terbium in Calcium Tungstate, Harry Diamond Laboratories, HDL-TR-1377.

Wortman, D. E., C. A. Morrison, and N. Karayianis (1977, June), Rare Earth Ion-Host Lattice Interactions: 5.--Lanthanides in $CaWO_4$, Harry Diamond Laboratories, HDL-TR-1794.

Wortman, D. E., and D. Sanders (1971a), Optical Spectrum of Trivalent Dysprosium in Calcium Tungstate, J. Chem. Phys. 55, 3212.

Wortman, D. E., and D. Sanders (1971b, April), Optical Spectrum of Trivalent Dysprosium in Calcium Tungstate, Harry Diamond Laboratories, HDL-TR-1540.

Wortman, D. E., and D. Sanders (1970a), Ground Term Energy Levels of Triply Ionized Holmium in Calcium Tungstate, J. Chem. Phys. 53, 1247.

Wortman, D. E., and D. Sanders (1970b, March), Absorption and Fluorescence Spectra of Ho^{3+} in $CaWO_4$, Harry Diamond Laboratories, HDL-TR-1480.

14. CRYSTAL-FIELD EFFECTS NOT YET FULLY INCORPORATED

14.1 Self-Consistent Point Dipole and Point Multipole

In section 12 we discussed the point-charge contribution to the multipolar field components A_{nm}. It was early recognized by Hutchings and Ray (1963) that the multipolar components of the constituent ions contribute to the A_{nm} at the site occupied by the unfilled shell nd^N. For a point charge eZ_i located at $\vec{R}_i$ from the origin ion site, we have the electric potential

$$\phi(\vec{r}) = \frac{eZ_i}{|\vec{R}_i - \vec{r}|} \quad . \tag{14.1}$$

The potential energy of one of the ℓ^N electrons at r is

$$U(\vec{r}) = -e\phi(r)$$

$$\tag{14.2}$$

$$= -e^2 Z_i \sum_{nm} \frac{r^n}{R_i^{n+1}} \, C_{nm}(\hat{r})c_{nm}^+(\hat{R}_i) \quad ,$$

where we have expanded the denominator of equation (14.1) in the spherical tensors discussed in chapter 1. If we write equation (14.2) as

$$U(\vec{r}) = \sum_{nm} A_{nm}^* r^n C_{nm}(r) \tag{14.3}$$

then

$$A_{nm}^{(0)} = -e^2 \sum_i \frac{Z_i C_{nm}(\hat{R}_i)}{R_i^{n+1}} \quad , \tag{14.4}$$

where the sum on i covers all the ions of charge eZ_i in the solid. This result we derived in section 11, expressed in slightly different form. It seems natural to extend equation (14.3) to the form

$$U(\vec{r}) = \sum_{n,m,k} A_{nm}^{(k)*} r^n C_{nm}(\hat{r}) \quad , \tag{14.5}$$

and relate the $A_{nm}^{(k)}$ to the various k-pole moments of ligands at $\vec{R}_i$. To relate the $A_{nm}^{(k)}$ to the multipole moments, eQ_{kq}, we need first to express the electric potential at the rare-earth electron due to the multipole moment $eQ_{kq}(i)$ at $\vec{R}_i$.

The electric potential due to a multipole distribution at $\vec{R}_i$ is given by

$$\phi(\vec{r}) = e \sum_{\substack{kq \\ nm}} (-1)^k Q_{kq}(i) \binom{2k+2n}{2n}^{1/2} \langle n(m)k(q)|n+k(m+q)\rangle \frac{C_{n+k,m+q}^*(\hat{R}_i)}{R_i^{n+k+1}} r^n C_{nm}(\hat{r})$$

$$\tag{14.6}$$

where

$$\binom{2k+2n}{2n} = \frac{(2k+2n)!}{(2n)!(2k)!}$$

(the details of the derivation of this result are given in sect. 15). Thus, since $U(\vec{r}) = -e\phi(\vec{r})$, we find, using equation (14.5), that

$$A_{nm}^{(k)} = -e^2 \sum_{q,i} (-1)^k Q_{kq}^*(i) \binom{2n+2k}{2n}^{1/2} \langle n(m)k(q)|n+k(m+q)\rangle \frac{C_{n+k,m+q}(\hat{R}_i)}{R_i^{n+k+1}} \quad . \tag{14.7}$$

If we let k = 0 in equation (14.7), we obtain

$$A_{km}^{(0)} = -e^2 \sum_i Q_{00}^*(i) \frac{C_{nm}(\hat{R}_i)}{R_i^{n+1}} \tag{14.8}$$

(if k = 0, q = 0), which is identical to the result given in equation (14.4) if we identify $Q_{00}^*(i)$ with Z_i (the number of charges) given there. The result for k = 1 is

$$A_{nm}^{(1)} = e^2 \sum_{q,i} Q_{1q}^*(i) \binom{2n+2}{2n}^{1/2} \langle n(m)1(q)|n+1(m+q)\rangle \frac{C_{n+1,m+q}(\hat{R}_i)}{R_i^{n+k+1}} .$$

(14.9)

Since

$$\langle n(m)1(q)|n+1(m+q)\rangle = (-1)^{1-q} \left(\frac{2n+3}{2n+1}\right)^{1/2} \langle 1(-q)n+1(m+q)|n(m)\rangle$$

and

$$Q_{1q}^* = (-1)^q Q_{1-q} ,$$

we can use these results in equation (14.9), to obtain

$$A_{nm}^{(1)} = -e^2 \sum_{q,i} \sqrt{(n+1)(2n+3)} \; Q_{1q}(i) \; \langle 1(q)n+1(m-q)|n(m)\rangle \frac{C_{n+1,m-q}(\hat{R}_i)}{R_i^{n+2}} ,$$

(14.10)

which is identical to the result published by Morrison (1976), if we identify $eQ_{1q}(i) = p_{1q}(i)$ (p_{1q} is the dipole moment component). Thus, if we knew the $Q_{kq}(i)$, we could easily calculate the $A_{nm}^{(k)}$ by using equation (14.7). Unfortunately, the real difficulty is determining the $Q_{kq}(i)$. In what follows we restrict our discussion to the dipole case, k = 1, and let $eQ_{1q} = p_q$ and express the results in Cartesian vectors.

At sites of low symmetry, an electric field can exist whose value is determined by the various point charges of the solid. The electric field due to the point charges of the solid at a site of low symmetry is given by

$$\vec{E}_j^0 = - \sum_i \frac{q_i \vec{R}_{ij}}{R_{ij}^3} ,$$

(14.11)

and the field generated by the point dipoles is

$$\vec{E}_j^d = \sum_i \left[\frac{3\vec{R}_{ij}(\vec{R}_{ij} \cdot \vec{p}_i)}{R_{ij}^5} - \frac{\vec{p}_i}{R_{ij}^3} \right] .$$

(14.12)

The dipole moment at site j is then given by

$$\vec{p}_j = \alpha\vec{E}_j = \alpha\left[\vec{E}_j^0 + \vec{E}_j^d\right] \quad , \tag{14.13}$$

where α is the polarizability of the ion at site j. (If more than one species is considered, the polarizability of each type must be used.) The sum in equation (14.11) presents no problem and can be done quite simply. To perform the sum in equation (14.12), it is convenient to assume a fixed coordinate system in the unit cell and an associated reference point (say position 1); then each dipole moment, $\vec{p}_j$, can be related to the dipole located at the reference moment, $\vec{p}_1$, by the symmetry operation of the crystal. Similarly, the field at each point, $\vec{E}_j$, can be related to $\vec{E}_1$. Having done this, we can write

$$\vec{E}_1^d = \underline{G}(1)\cdot\vec{p}_1 \quad , \tag{14.14}$$

and from equation (14.13)

$$\vec{p}_1 = \alpha\left[\vec{E}_1^0 + \underline{G}(1)\cdot\vec{p}_1\right] \quad . \tag{14.15}$$

The result given in equation (14.15) can then be solved for the dipole moment $\vec{p}_1$ to give

$$\vec{p}_1 = \alpha\underline{B}(1)\cdot\vec{E}_1^0 \quad , \tag{14.16}$$

where

$$\underline{B} = (1 - \alpha\underline{G})^{-1} \quad .$$

The result obtained in equation (14.16) is rather interesting; if the polarizability, α, is near the reciprocal of one of the eigenvalues of the $\underline{G}$ matrix, then the dipole moment becomes excessively large. This is suggestive of the type of catastrophe that occurs in the onset of a ferroelectric transition. Such a situation would, perhaps, be modified by the inclusion of the higher multipole moments in the calculation. It should be pointed out that the $\underline{G}$ matrix defined in equations (14.12) and (14.15) is dependent only on the lattice constants and the symmetry of the crystal. The results here were expressed in terms of Cartesian coordinates but can equally well be done in spherical tensors. If higher moments were considered, the spherical tensor form would be much more convenient. (This statement has been confirmed by M. Faucher in private communication, 1982. She has extended the self-consistent moments through quadrupole moments.)

14.2 Self-Consistent Results for Scheelite Structure

The procedure given above is rather involved, so we shall go into the derivation of the $\underline{G}$ tensor for the scheelite structure ($CaWO_4$, $LiYF_4$, etc). The space group for scheelite is 88 in the International Tables; the position of all the constituents is given in table 14.1. To be specific, $LiYF_4$ has been chosen; the fluorine is in site 1; x, y, and z have been chosen as the reference points for the dipoles (u, v, w); and all other dipoles in the unit cell are related to u, v, and w. No dipoles can exist at the Y or Li sites since the lowest fields at these sites are quadrupole (k = 2).

To evaluate $\underline{G}$ for the scheelite structure, we choose the ion at site 1 in table 14.1 as j in equation (14.12). The $\vec{R}$ in equation (14.5), including the translational vectors (ℓ in x, m in y, n in z), is

$$\vec{R}_{i,1} = (\ell+x_i-x)a\hat{e}_x + (m+y_i-y)a\hat{e}_y + (n+z_i-z)c\hat{e}_x \quad , \qquad (14.17)$$

where we shall, during this discussion, suppress the explicit dependence of $\vec{R}$ on ℓ, m, and n. We write equation (14.12) as

$$\vec{E}_1^d = \vec{F}_1^d + \vec{I}_1^d \quad , \qquad (14.18)$$

where

$$\vec{F}_1^d = - \sum_i \frac{\vec{p}_1}{R_{i,1}^3} \qquad (14.19)$$

and

$$\vec{I}_1^d = 3 \sum_i \vec{R}_{i,1} \frac{\vec{R}_{i1}\cdot\vec{p}_1}{R_{i,1}^5} \quad , \qquad (14.20)$$

where sums over ℓ, m, and n are implicit. Then using table 14.1, we write $\vec{F}_1^d$ explicitly as

$$F_x^d = - \frac{u}{R_{1,1}^3} - \frac{v}{R_{2,1}^3} + \frac{u}{R_{3,1}^3} + \frac{v}{R_{4,1}^3} - \frac{u}{R_{5,1}^3} - \frac{v}{R_{6,1}^3} + \frac{u}{R_{7,1}^3}$$

$$(14.21)$$

$$+ \frac{v}{R_{8,1}^3} - \frac{u}{R_{9,1}^3} - \frac{v}{R_{10,1}^3} + \frac{u}{R_{11,1}^3} + \frac{v}{R_{12,1}^3} - \frac{u}{R_{13,1}^3} - \frac{v}{R_{14,1}^3} + \frac{U}{R_{15,1}^3} + \frac{v}{R_{16,1}^3} \quad .$$

TABLE 14.1. SPACE GROUP 88 (FIRST SETTING): COORDINATES OF ALL
IONS IN A UNIT CELL OF $YLiF_4$ AND DIPOLE MOMENTS OF EACH ION
(p_x, p_y, and p_z of site 1 are chosen as u, v, and w, respectively)

Site	Ion	x	y	z	p_x	p_y	p_z	Q_{kq}[a]
1	F	x	y	z	u	v	w	1
2	F	y	-x	-z	v	-u	-w	$(-1)^k(i)^q$
3	F	-x	-y	z	-u	-v	w	$(-1)^q$
4	F	-y	x	-z	-v	u	-w	$(-1)^k(-i)^q$
5	F	1/2 + x	1/2 + y	1/2 + z	u	v	w	1
6	F	1/2 + y	1/2 - x	1/2 - z	v	-u	-w	$(-1)^k(i)^q$
7	F	1/2 - x	1/2 - x	1/2 - z	-u	-v	w	$(-1)^q$
8	F	1/2 - y	1/2 - y	1/2 + z	-v	u	-w	$(-1)^k(-i)^q$
9	F	x	1/2 + y	1/4 - z	u	v	-w	$(-1)^{k+q}$
10	F	y	1/2 - x	1/4 + z	v	-u	w	$(-i)^q$
11	F	-x	1/2 - y	1/4 - z	-u	-v	-w	$(-1)^k$
12	F	-y	1/2 + x	1/4 + z	-v	u	w	$(i)^q$
13	F	1/2 + x	y	3/4 - z	u	v	-w	$(-1)^{k+q}$
14	F	1/2 + y	-x	3/4 + z	v	-u	w	$(-i)^q$
15	F	1/2 - x	-y	3/4 - z	-u	-v	-w	$(-1)^k$
16	F	1/2 - y	x	3/4 + z	-v	u	w	$(i)^q$
17	Li	0	0	0	--	--	--	--
18	Li	0	1/2	1/4	--	--	--	--
19	Li	1/2	1/2	1/2	--	--	--	--
20	Li	1/2	0	3/4	--	--	--	--
21	Y	0	0	1/2	--	--	--	--
22	Y	1/2	0	1/4	--	--	--	--
23	Y	1/2	1/2	0	--	--	--	--
24	Y	0	0	3/4	--	--	--	--

[a]The last column relates those Q_{kq} to the reference point Q_{kq}.
Thus the Q_{kq} for fluorine are all related to site 1.

Then we can write

$$F_x^d = \sum_{i=1}^{8} \frac{(-1)^i}{R_{2i+1,1}^3} u + \sum_{i=1}^{8} \frac{(-1)^i}{R_{2i,1}^3} v + (0)w \quad . \tag{14.22}$$

If we let $\underline{G} = \underline{G}' + 3\underline{G}''$ and restore the ℓ, m, n sum, we have

$$G'_{xx} = \sum_{\ell,m,n} \sum_{i=1}^{8} \frac{(-1)^i}{R_{2i+1,1}^3} \quad , \tag{14.23}$$

$$G'_{xy} = \sum_{\ell,m,n} \sum_{i=1}^{8} \frac{(-1)^i}{R_{2i,1}^3} \quad , \tag{14.24}$$

$$G'_{xz} = 0 \quad . \tag{14.25}$$

By similar methods we obtain

$$G'_{zz} = \sum_{\ell,m,n} \sum_{i=1}^{8} (-1)^i \left[\frac{1}{R_i^3} - \frac{1}{R_{i+8}^3} \right] \quad , \tag{14.26}$$

and the $\underline{G}'$ tensor is symmetrical.

To evaluate the $\underline{G}''$ term, the procedure is precisely the same as to evaluate the $\underline{G}'$ term, except that we relate $\underline{G}''$ to equation (14.20). It is convenient to express $\vec{R}_{i1} \cdot \vec{P}_i$ explicitly in tabular form, as given in table 14.2, for easy reference when writing out each term of $\underline{G}''$. We shall not write out the detailed expression as in equation (14.21), but this procedure gives

$$G''_{xx} = - \sum_{i=1}^{8} (-1)^i [X_{2i+1}^2 - X_{2i}Y_{2i}] \quad , \tag{14.27}$$

$$G''_{xy} = - \sum_{i=1}^{8} (-1)^i [X_{2i}^2 - X_{2i}Y_{2i-1}] \quad , \tag{14.28}$$

$$G''_{xz} = - \sum_{i=1}^{8} (-1)^i [X_i Z_i - X_{i+8}Z_{i+8}] \quad , \tag{14.29}$$

$$G''_{yz} = - \sum_{i=1}^{8} (-1)^i [Y_i Z_i - Y_{i+8}Z_{i+8}] \quad , \tag{14.30}$$

$$G''_{zz} = - \sum_{i=1}^{8} (-1)^i [Z_i^2 - Z_{i+8}^2] \quad , \tag{14.31}$$

134

where

$$X_i Y_i = \frac{a^2 x(i) y(i)}{R_{1,1}^5} \quad ,$$

$$X_i^2 = \frac{a^2 x(i)^2}{R_{1,1}^5} \quad ,$$

$$Y_i^2 = \frac{a^2 y(i)^2}{R_{1,1}^5} \quad ,$$

$$X_i Z_i = \frac{acx(i)z(i)}{R_{1,1}^5} \quad ,$$

and all the sums in equations (14.27) through (14.31) have the sum over ℓ, m, and n implied. The $\underline{G}''$ is symmetric (this can be shown directly from evaluating, for example, G''_{xy} and G''_{yx} independently).

TABLE 14.2. VALUES OF $\vec{R} \cdot \vec{P}$ FOR DIFFERENT SITES IN SCHEELITE[a]

Site	p_x	p_y	p_z	$\vec{R} \cdot \vec{P}$
1	u	v	w	$x(1)u + y(1)v + z(1)w$
2	v	-u	-w	$x(2)v - y(2)u - z(2)w$
3	-u	-v	w	$-x(3)u - y(3)v + z(3)w$
4	-v	u	-w	$-x(4)v + y(4)u - z(4)w$
5	u	v	w	$x(5)u + y(5)v + z(5)w$
6	v	-u	-w	$x(6)v - y(6)u - z(6)w$
7	u	-v	w	$-x(7)u - y(7)v + z(7)w$
8	-v	u	-w	$-x(8)v + y(8)u - z(8)w$
9	u	v	-w	$x(9)y + y(9)v - z(9)w$
10	v	-u	w	$x(10)v - y(10)u + z(10)w$
11	-u	-v	-w	$-x(11)u - y(11)v - z(11)w$
12	-v	u	w	$-x(12)v + y(12)u - z(12)w$
13	u	v	-w	$x(13)u + y(13)v - z(13)w$
14	v	-u	w	$x(14)v - y(14)u + z(14)w$
15	-u	-v	-w	$-x(15)u - y(15)v - z(15)w$
16	-v	u	w	$-x(16)v - y(16)u - z(16)w$

[a] $x(i) = \ell + x_i - x_1$, $y(i) = m + y_i - y_1$, $z(i) = n + z_i - z_1$.

The equations for $\underline{G}'$ and $\underline{G}''$ were calculated for several lattices, and the results are given in table 14.3. The crystal axial field components, A_{no}, were computed for $CaWO_4$ using $\alpha = 2.4$ $\text{\AA}^3$ and oxygen charge of $-2e$, and using $\alpha = 0.24$ $\text{\AA}^3$ and oxygen charge of $-e$. The results are shown in table 14.4 (Morrison, 1976).

After the above work had been done, the dipole terms in the A_{nm}^d were programmed for a computer for all the 230 space groups. In the

TABLE 14.3. $\underline{G}$ TENSOR AND X-RAY DATA FOR SEVERAL COMPOUNDS $(1/\text{\AA}^3)$[a,b]

Compound	a	c	x	y	z
$CaWO_4$	5.248	11.376	0.2413	0.1511	0.0861
$PbMoO_4$	5.4312	12.0165	0.2353	0.13660	0.08110
$YLiF_4$	5.1668	10.733	0.2820	0.1642	0.0815
YVO_4	7.120	6.289	0.1852	0	0.1749

	G_{xx}	G_{xy}	G_{xz}	G_{yy}	G_{yz}	G_{zz}
$CaWO_4$	-0.252608	-0.0731076	-0.0979719	0.197101	-0.040229	0.123419
$PbMoO_4$	-0.224152	-0.0758124	-0.0976969	0.168517	-0.0426228	0.0768659
$YLiF_4$	-0.252131	-0.0826652	-0.125165	0.192571	-0.0446824	0.168842
YVO_4	-0.734174	0	-0.241965	0.127677	0	-0.173784

[a] The reference site 1, in all the calculations, is the ligand at x, y, and z.
[b] YVO_4 is not a scheelite structure (YVO_4 is the zircon structure, space group 141, in the International Tables) but can be done in the scheelite structure by translating the oxygen positions to the above (see Karayianis and Morrison, 1973).

TABLE 14.4. AXIAL COMPONENTS OF CRYSTAL FIELD FOR TWO
VALUES OF POLARIZABILITY OF OXYGEN AND OXYGEN CHARGE
(cm^{-1}/A^3)

Component	$\alpha = 2.4\ A^3,\ q_{0x} = -2e$		$\alpha = 0.24\ A^3,\ q_{0x} = -e$	
	A_{n0}^o	A_{n0}^d	A_{n0}^o	A_{n0}^d
A_{20}	10,115	-22,954	2321.1	-692.58
A_{40}	-4,215.4	14,975	-1919.6	332.25
A_{60}	38.625	-897.94	7.2194	-10.206

program any number of inequivalent sites can have an associated dipole moment (we only considered one type of site above). Recently the members of Caro's group in France and de Sa's group in Brazil (Faucher and Malta, 1981) have included the dipole and quadrupole moments in a self-consistent manner for LaCl$_3$; they have found that with the reported values of the dipole and quadrupole polarizabilities the resultant A_{nm}^q is much larger than A_{nm}^o or A_{nm}^d.

14.3 Self-Induced Effects

When a rare-earth ion is immersed in a solid it is possible for its electrons to experience a field due to the reaction of the medium back on the electrons. Both this type of field and the external fields due to the point charges of the medium can exist. This reaction is identical to the classical problem of a charged particle interacting with its induced image in a conducting plate or sphere. The interaction was recognized by Judd (1977), and it was he who suggested the polarization of the ligands as a possible source of a two-electron crystal-field interaction. In this section we consider the development of this interaction as derived earlier (Morrison, 1980), using the same technique used in the earlier work. In later sections this interaction is developed in a more general way, deriving the multipolar interaction.

We consider an electron at $\vec{r}$ on a rare-earth ion and a ligand at $\vec{R}$ with polarizability α. The electric potential created by the electron is

$$\phi = \frac{-e}{|\vec{R} - \vec{r}|} \ . \tag{14.32}$$

The electric field at the ligand is

$$\vec{E} = -\nabla_R \phi \ ,$$

where ∇_R indicates that the derivative should be taken with respect to $\vec{R}$. Then,

$$\vec{E} = \frac{-e(\vec{R} - \vec{r})}{|\vec{R} - \vec{r}|^3} \ . \tag{14.33}$$

136

The dipole moment induced on the ligand is given by

$$\vec{p} = \alpha \vec{E} \quad , \tag{14.34}$$

where α is the polarizability of the ligand.

Now if we consider a dipole from some arbitrary origin, the electric potential at point $\vec{R}_1$ *from* that origin is

$$\phi_1 = \frac{\vec{p} \cdot \vec{R}_1}{R_1^3} \quad . \tag{14.35}$$

To find this potential at the electron itself, we let $\vec{R}_1 = -(\vec{R} - \vec{r})$. Then equation (14.35) becomes

$$\phi_1 = \frac{-\vec{p} \cdot (\vec{R} - \vec{r})}{|\vec{R} - \vec{r}|^3} \quad . \tag{14.36}$$

The energy of the electron interacting with this potential is given by

$$U(\vec{r}, \vec{R}) = \frac{-e}{2} \phi_1(r)$$

$$\tag{14.37}$$

$$= \frac{e}{2} \frac{\vec{p} \cdot (\vec{R} - \vec{r})}{|\vec{R} - \vec{r}|^3} \quad ,$$

where the 1/2 is due to a self-interaction. We can write

$$\frac{\vec{R} - \vec{r}}{|\vec{R} - \vec{r}|^3} = -\nabla_R \frac{1}{|\vec{R} - \vec{r}|} \quad . \tag{14.38}$$

Then equation (14.37) becomes

$$U(\vec{r}, \vec{R}) = \frac{-e}{2} \vec{p} \cdot \nabla_R \frac{1}{|\vec{R} - \vec{r}|} \quad , \tag{14.39}$$

and similarly

$$\vec{E} = e \nabla_R \frac{1}{|\vec{R} - \vec{r}|} \quad . \tag{14.40}$$

Using the result of equation (14.40) in equation (14.34) and substituting the result into equation (14.39), we have

$$U(\vec{r},\vec{R}) = \frac{-e^2}{2} \alpha \left(\nabla_R \frac{1}{|\vec{R} - \vec{r}|} \right) \cdot \left(\nabla_R \frac{1}{|\vec{R} - \vec{r}|} \right) \quad , \tag{14.41}$$

where ∇_R operates only on the function on its immediate right. To further reduce the result given in equation (14.41), we consider the operation

$$\nabla^2(\psi_1\psi_2) = \psi_1 \nabla^2 \psi_2 + \psi_2 \nabla^2 \psi_1 + 2(\nabla\psi_1) \cdot (\nabla\psi_2) \quad . \tag{14.42}$$

If ψ_1 and ψ_2 satisfy Laplace's equation (which they do), then

$$\nabla^2(\psi_1\psi_2) = 2(\nabla\psi_1) \cdot (\nabla\psi_2) \quad . \tag{14.43}$$

If we identify ψ_1 and ψ_2 with $1/|\vec{R} - \vec{r}|$ in equations (14.41) and (14.43), we can write $U(\vec{r},\vec{R})$ as

$$U(\vec{r},\vec{R}) = \frac{-\alpha e^2}{4} \nabla_R^2 \frac{1}{|\vec{R} - \vec{r}|^2} \quad . \tag{14.44}$$

To proceed further we must expand the factors on the right side of equation (14.44). First we notice that

$$|\vec{R} - \vec{r}|^2 = R^2 + r^2 - 2\vec{r}\cdot\vec{R}$$

$$= 2Rr \left[\frac{R^2 + r^2}{2Rr} - \hat{r}\cdot\hat{R} \right] \quad . \tag{14.45}$$

If we let

$$t = \frac{R^2 + r^2}{2rR} \quad , \tag{14.46}$$

then

$$\frac{1}{|\vec{R} - \vec{r}|^2} = \frac{1}{2rR} \left(\frac{1}{t - z} \right) \tag{14.47}$$

with $z = \hat{r}\cdot\hat{R}$.

The expansion

$$\frac{1}{t - z} = \sum_n (2n + 1)Q_n(t)P_n(z) \tag{14.48}$$

is given by Rainville (1960); the leading term for large t is

$$Q_n(t) \simeq \frac{2^n(n!)^2}{t^{n+1}(2n + 1)!} \quad . \tag{14.49}$$

From equation (14.46) we have

$$Q_n(t) \simeq \frac{2^{2n+1}(n!)^2}{(2n + 1)!} \frac{r^{n+1}}{R^{n+1}} \tag{14.50}$$

for large R.

Substituting the result of equation (14.50) into equation (14.47) gives

$$\frac{1}{|R - r|^2} = \sum \frac{2^{2n}(n!)^2 r^n}{(2n)!R^{n+2}} P_n(z) \quad . \tag{14.51}$$

From the Legendre addition theorem (see sect. 1), we have

$$P_n(z) = P_n(\hat{r} \cdot \hat{R})$$

$$\tag{14.52}$$

$$= \sum_{n,m} C_{nm}(\hat{r})C^*_{nm}(\hat{R})$$

and

$$\frac{1}{|R - r|^2} = \sum_{n,m} \frac{2^{2n}(n!)^2}{(2n)!} \frac{r^n}{R^{n+2}} C_{nm}(\hat{r})C^*_{nm}(\hat{R}) \quad . \tag{14.53}$$

The remaining necessary operation is ∇^2_R, which can be written

$$\nabla^2_R = \frac{1}{R^2} \frac{d}{dR} \left(R^2 \frac{d}{dR}\right) - \frac{\vec{L}^2}{R^2} \quad . \tag{14.54}$$

The only term in equation (14.53) that this operates on is

$$\frac{1}{R^{n+2}} C^*_{nm}(\hat{R}) \quad . \tag{14.55}$$

139

Using equations (14.54) and (14.55), we have

$$\nabla_R^2 \left[R^{-n-2} C_{nm}^* (\hat{R}) \right] = \left[\frac{(n+1)(n+2)}{R^{n+4}} - \frac{n(n+1)}{R^{n+4}} \right] C_{nm}^* (\hat{R}) \quad . \qquad (14.56)$$

Finally,

$$\nabla_R^2 \left[\frac{1}{R^{n+2}} C_{nm}^* (\hat{R}) \right] = \frac{2(n+1)}{R^{n+4}} C_{nm}^* (\hat{R}) \quad , \qquad (14.57)$$

where we have used

$$(\vec{L})^2 C_{nm} (\hat{R}) = n(n+1) C_{nm} (\hat{R}) \quad , \qquad (14.58)$$

a result we discussed in section 1. The result in equation (14.57) substituted into equation (14.53) gives

$$\nabla_R^2 \frac{1}{|R-r|^2} = \sum_{nm} \frac{2^{2n+1}(n!)(n+1)!}{(2n)!} \frac{r^n}{R^{n+4}} C_{nm}(\hat{r}) C_{nm}^*(\hat{R}) \quad . \qquad (14.59)$$

The result given in equation (14.59) is substituted into equation (14.44). This result, when summed over all ligands at $\vec{R}_j$ with polarizability α_j, produces

$$U(r) = \frac{-e^2}{4} \sum_{\substack{nm \\ j}} \alpha_j \left[\frac{2^{2n+1} n! (n+1)!}{(2n)!} \frac{C_{nm}^* (\hat{R}_j)}{R_j^{n+4}} \right] r^n C_{nm}(\hat{r}) \quad . \qquad (14.60)$$

If we write equation (14.60) as we have previously done with the point-charge model,

$$U(r) = \sum_{nm} \left(A_{nm}^{SI} \right)^* r^n C_{nm}(\hat{r}) \quad , \qquad (14.61)$$

we have

$$A_{nm}^{SI} = \left(\frac{e^2}{4} \right) \frac{2^{2n+1} n! (n+1)!}{(2n)!} \sum_j \frac{\alpha_j C_{nm}(\hat{R}_j)}{R_j^{n+4}} \quad , \qquad (14.62)$$

which are the self-induced crystal-field components due to induced dipoles only. Higher order multipole moments can be induced on the ligands, and these multipoles will contribute a correction. From previous experience, we should anticipate the total self-induced multipole fields to be of the form

$$A_{nm}^{SI} = \sum_{k=1}^{\infty} A_{nm}^{SI}(k) \qquad (14.63)$$

with the result above being $A_{nm}^{SI}(1)$.

As in the point-charge model, if we express all lengths in angstroms and α_j in angstroms cubed, then equation (14.62) becomes

$$A_{nm}^{SI}(1) = -\left(\frac{\alpha_o}{8\pi} \times 10^8\right) \frac{2^{2n+1} n!(n+1)!}{(2n)!} \sum_j \frac{\alpha_j C_{nm}(\hat{R}_j)}{R_j^{n+4}} \quad . \tag{14.64}$$

To express A_{nm}^{SI} in units of cm^{-1}/A^n, use the conversion factor $\alpha_o/8\pi \times 10^8 = 29,035$.

14.4 Annotated Bibliography and References

Faucher, M., and P. Caro (1977), A Quickly Converging Method for Computing Electrostatic Crystal Field Parameters, J. Chem. Phys. 66, 1273.

Faucher, M., and D. Garcia (1983a), Crystal Field Effects in Rare-Earth-Doped Oxyhalides: Ab-Initio Calculations Including Effects of Dipolar and Quadrupolar Moments, Solid State Chemistry, Proceedings of the Second European Conference (7-9 June 1982, Veldhoven, Netherlands), Elsevier, Amsterdam, Netherlands, p 547.

Faucher, M., and D. Garcia (1983b), Crystal Field Effects on 4f Electrons: Theories and Reality, J. Less-Common Metals 93, 31.

Faucher, M., and D. Garcia (1982), Electrostatic Crystal-Field Contributions in Rare-Earth Compounds with Consistent Multipolar Effects: I.--Contribution to K-Even Parameters, Phys. Rev. B26, 5451.

Garcia, D. (1983), Simulation ab-initio des parametres du champ des ligandes, thesis, L'Ecole Centrale des Arts et Manufactures.

Garcia, D., and M. Faucher (1984), Crystal-Field Parameters in Rare-Earth Compounds: Extended Charge Contributions, Phys. Rev. A30, 1703.

Garcia, D., M. Faucher, and O. L. Malta (1983), Electrostatic Crystal-Field Contribution in Rare-Earth Compounds with Consistent Multipolar Effects: II.--Contribution to K-Odd Parameters (Transition Probabilities), Phys. Rev. B27, 7386.

Hutchings, M. T., and D. K. Ray (1963), Investigations into the Origin of Crystalline Electronic Field Effects on Rare-Earth Ions: I--Contributions from Neighbouring Induced Moments, Proc. Phys. Soc. 81, 663.

International Tables for X-Ray Crystallography (1952), Kynock Press, U.K.

Jackson, J. D. (1975), Classical Electrodymanics, Wiley, New York, NY, chapters 3 and 4. Note that his multipole $q_{\ell m}$ are related to our Q_{kq} by $q^*_{\ell m} = \left(\sqrt{(2\ell+1)/4\pi}\right)Q_{\ell m}$.

Judd, B. R. (1977), Correlation Crystal Fields for Lanthanide Ions, Phys. Rev. Lett. 39, 242.

Morrison, C. A. (1980, January 15), Host Dependence of the Rare-Earth Ion Energy Separation $4F^N - 4F^{N-1}n\ell$, J. Chem. Phys. 72, 1001.

Morrison, C. A. (1976), Dipolar Contributions to the Crystal Fields in Ionic Solids, Solid State Commun. 18, 153.

Morrison, C. A., G. F. de Sa, and R. P. Leavitt (1982), Self-Induced Multipole Contribution to the Single-Electron Crystal Field, J. Chem. Phys. 76, 3899.

Rainville, E. D. (1960), Special Functions, Macmillan, New York, NY, p 1822.

15. MISCELLANEOUS CRYSTAL-FIELD EFFECTS

15.1 Judd's Interaction for Two Electrons

The interaction considered here is a development of a suggestion by Judd (1977) concerning a possible origin of two-electron crystal-field effects. Specifically, Judd suggested that such terms would arise if one of the electrons in the configuration $n\ell^N$ polarized a nearby ion, and the remaining $N - 1$ electrons interacted with the induced multipolar moments. The investigation of this interaction was performed later (Morrison, 1980), assuming only a dipole polarizability. The interaction for two electrons that resulted is

$$V(1,2,R) = \sum_{\substack{a,\alpha \\ b,k,q}} F(abk) r_1^a C_{a\alpha}(\hat{r}_1) r_2^b C_{b,q-\alpha}(\hat{r}_2) \langle a(\alpha)b(q-\alpha)|k(q)\rangle \frac{C_{kq}^*(\hat{R})}{R^{a+b+4}} \quad (15.1)$$

where

$$F(abk) = -\left(\alpha \frac{e^2}{2}\right) \langle a(0)b(0)|k(0)\rangle \left[(a+b+1)(a+b+2) - k(k+1)\right] \quad,$$

and α is the dipole polarizability of the ion at R.

The development of the result given in equation (15.1) was similar to that given in the derivation of the self-induced field in section 14.3. For the full multipolar result we shall use more general methods.

The electric potential of an electron at $\vec{r}_1$, as seen at a ligand at $\vec{R}$, is

$$\phi(\hat{R}_1) = \frac{-e}{|\vec{R}_1|} \quad , \quad (15.2)$$

where

$$\vec{R}_1 = \vec{R} - \vec{r}_1 \quad .$$

The multipolar inducing field E_{nm} at $\vec{R}_1$ can be defined by

$$\phi(\vec{R}_1 + \vec{x}) = - \sum_{nm} E^*_{nm} x^n C_{nm}(\hat{x}) \quad . \tag{15.3}$$

By expanding equation (15.2), we obtain

$$\phi(\vec{R}_1 + \vec{x}) = -e \sum_{nm} (-1)^n \frac{C^*_{nm}(\hat{R}_1) x^n}{R_1^{n+1}} C_{nm}(\hat{x}) \quad ; \tag{15.4}$$

then comparing equation (15.4) with equation (15.3) gives

$$E_{nm} = e(-1)^n C_{nm}(\hat{R}_1)/R_1^{n+1} \quad . \tag{15.5}$$

The multipole moment, Q_{nm}, is given by

$$Q_{nm} = \alpha_n E_{nm} \quad , \tag{15.6}$$

where the multipole polarizability is α_n.

The electric potential at an arbitrary point $\vec{R}_3$ from a multipole distribution is given by

$$\phi_Q(\vec{R}_3) = \sum \frac{Q^*_{nm} C_{nm}(\hat{R}_3)}{R_3^{n+1}} \quad , \tag{15.7}$$

and the energy of an electron at $\vec{r}_2$ interacting with the multipoles is

$$U = -e\phi_Q \qquad (\vec{R}_3 = -\vec{R}_2) \tag{15.8}$$

with $\vec{R}_2 = \vec{R} - \vec{r}_2$. From equation (15.7), we obtain

$$U = -e \sum_{nm} \frac{Q^*_{nm} (-1)^n C_{nm}(\hat{R}_2)}{R_2^{n+1}} \quad . \tag{15.9}$$

Now from equations (15.6) and (15.5) we have

$$Q_{nm} = e\alpha_n (-1)^n \frac{C_{nm}(\hat{R}_1)}{R_1^{n+1}} \quad , \tag{15.10}$$

which, when substituted into equation (15.9), gives

$$U(\vec{r}_1 \vec{r}_2, \vec{R}) = -e^2 \sum_{nm} \alpha_n \frac{C_{nm}^*(\hat{R}_1)}{R_1^{n+1}} \frac{C_{nm}(\hat{R}_2)}{R_2^{n+1}} \quad . \tag{15.11}$$

If we write

$$U(\vec{r}_1 \vec{r}_2, \vec{R}) = \sum_{nm} U^{(n)}(\vec{r}_1 \vec{r}_2, \vec{R}) \quad , \tag{15.12}$$

we have

$$U^{(n)}(\vec{r}_1 \vec{r}_2, \vec{R}) = -e^2 \alpha_n \sum_m \frac{C_{nm}^*(\hat{R}_1)}{R_1^{n+1}} \frac{C_{nm}(\hat{R}_2)}{R_2^{n+1}} \quad . \tag{15.13}$$

If we were considering the self-interaction, at this point we would let $\vec{R}_2 = \vec{R}_1$ and take half the results. The sum on m would then collapse to unity and $U^{(n)}(\vec{r}_1 \vec{r}_1, \vec{R}) = -e^2 (\alpha_n/2)/R_1^{2n+2}$.

However, the two-electron interaction is more complicated. We use the two-center expansions (Carlson and Rushbrooke, 1950; Judd, 1975) to obtain

$$\frac{C_{nm}^*(\hat{R}_1)}{R_1^{n+1}} = \sum_{a\alpha} \binom{2a+2n}{2a}^{1/2} \langle a(\alpha)n(m)|a+n(\alpha+m)\rangle \frac{r_1^a C_{a\alpha}(\hat{r}_1)}{R^{a+n+1}} C_{a+n,\alpha+m}^*(\hat{R}) \quad , \tag{15.14}$$

where $\vec{R}_1 = \vec{R} - \vec{r}_1$, and

$$\frac{C_{nm}(\hat{R}_2)}{R_2^{n+1}} = \sum_{b\beta} \binom{2b+2n}{2b}^{1/2} \langle b(\beta)n(m)|b+n(\beta+m)\rangle \frac{r_2^b C_{b\beta}^*(\hat{r}_2)}{R^{b+n+1}} C_{b+n,\beta+m}(\hat{R}) \quad , \tag{15.15}$$

where $\vec{R}_2 = \vec{R} - \vec{r}_2$.

As indicated in equation (15.11), equations (15.14) and (15.15) are to be multiplied together. When these two equations are multiplied, the two spherical tensors in $\hat{R}$ can be recoupled as

$$C_{a+n,\alpha+m}^*(\hat{R}) C_{b+n,\beta+m}(\hat{R}) = (-1)^{\alpha+m} \sum_k \langle b+n(0)a+n(0)|k(0)\rangle \tag{15.16}$$

$$\times \langle b+n(\beta+m)a+n(-\alpha-m)|k(\beta-\alpha)\rangle C_{k,\beta-\alpha}(\hat{R}) \quad ,$$

145

where we have used

$$C^*_{a+n,\alpha+m}(\hat{R}) = (-1)^{\alpha+m} C_{a+n,-\alpha-m}(\hat{R}) \quad . \tag{15.17}$$

It should be noted that the resultant projection in equation (15.16), $[C_{k,\beta-\alpha}(\hat{R})]$, is independent of m. Thus with a proper recoupling of the C-G coefficients in equations (15.14) and (15.15), the sum over m can be performed. Selecting the independent terms from the product of equations (15.14) and (15.15) and the result of equation (15.16), we have

$$S = \sum_m (-1)^{m+\alpha} \langle a(\alpha)n(m)|a+n(\alpha+m)\rangle \langle n(m)b(\beta)|b+n(\beta+m)\rangle \tag{15.18}$$

$$\times \langle b+n(\beta+m)a+n(-\alpha-m)|k(\beta-\alpha)\rangle \quad ,$$

which, when further reduced, gives

$$U^{(n)}(\vec{r}_1,\vec{r}_2,\vec{R}) = -e^2 \alpha_n \sum_{\substack{a\alpha \\ b\beta}} \sum_k \langle b+n(0)a+n(0)|k(0)\rangle \left[\binom{2a+2n}{2a}\binom{2b+2n}{2b}\right]^{1/2} \tag{15.19}$$

$$\times r_1^a C_{a\alpha}(\hat{r}_1) r_2^b C^*_{b\beta}(\hat{r}_2) \frac{C_{k,\beta-\alpha}(\hat{R})S}{R^{a+b+2n+2}} \quad .$$

Thus the final desired result is obtained if we know S. In equation (15.18) we rearrange the C-G coefficients as follows:

$$\langle a(\alpha)n(m)|a+n(a+m)\rangle = (-1)^{a-\alpha}\left(\frac{2a+2n+1}{2n+1}\right)^{1/2} \langle a(-\alpha)a+n(\alpha+m)|n(m)\rangle \quad , \tag{15.20}$$

$$\langle n(m)b(\beta)|b+n(\beta+m)\rangle = (-1)^{n-m}\left(\frac{2b+2n+1}{2b+1}\right)^{1/2} \langle n(m)b+n(-\beta-m)|b(-\beta)\rangle \quad .$$

We then recouple (eq (5.8)) the two C-G coefficients on the right to give

$$\langle a(-\alpha)a+n(\alpha+m)|n(m)\rangle \langle n(m)b+n(-\beta-m)|b(-\beta)\rangle$$

$$= \sum_f \sqrt{(2f+1)(2n+1)}\; W(a,a+n,b,b+n;nf)\; \langle a+n(\alpha+m)b+n(-\beta-m)|f(\alpha-\beta)\rangle \tag{15.21}$$

$$\times \langle a(-\alpha)f(\alpha-\beta)|b(-\beta)\rangle \quad .$$

The sum on m can now be performed; note that the phase, $(-1)^m$, in equation (15.20) cancels the $(-1)^m$ in equation (15.19) if we change the phase in the first C-G coefficient on the right side of equation (15.21). This then fixes the sums on f at k. Thus,

$$S = (-1)^{k-b+n}\left[\frac{(2a+2n+1)(2b+2n+1)}{2b+1}\right]^{1/2}\sqrt{2k+1}$$

(15.22)

$$\times\ W(a,a+n,b,bn;nk)\ \langle a(-\alpha)k(\alpha-\beta)|b(-\beta)\rangle\ .$$

The C-G in equation (15.22) can be rearranged to give

$$S = (-1)^{\alpha}\left[(2a+2n+1)(2b+2n+1)\right]^{1/2}W(a,a+n,b,b+n;nk)\ \langle a(\alpha)b(\beta)|k(\alpha-\beta)\rangle\ .$$

(15.23)

If we let

$$F_n(abk) = -\left(\alpha_n e^2\right)\langle a+n(0)b+n(0)|k(0)\rangle\ \sqrt{(2a+2n+1)(2b+2n+1)}$$

(15.24)

$$\times\ W(a,a+n,b,b+n;nk)\left[\binom{2a+2n}{2a}\binom{2b+2n}{2b}\right]^{1/2}\ ,$$

then, substituting into equation (15.19), we have

$$U^{(n)}(\vec{r}_1\vec{r}_2,\vec{R}) = \sum_{\substack{a,b\\k,q}} F_n(abk)r_1^a r_2^b \sum_{\alpha}\langle a(\alpha)b(q-\alpha)|k(q)\rangle\ C_{a\alpha}(\hat{r}_1)C_{b,q-\alpha}(\hat{r}_2)$$

(15.25)

$$\times\ \frac{C^*_{kq}(\hat{R})}{R^{q+b+2n+2}}\ ,$$

which is the final form of the two-electron multipolar interaction. To obtain the result given in (15.1), we would have to relate $\langle a+1(0)b+1(0)|k(0)\rangle$ to $\langle a(0)b(0)|k(0)\rangle$ and evaluate $W(a,a+1,b,b+1;1k)$, both of which procedures can be found in Rose (1957, pp 47, 277). If this is done, then equation (15.25) will reduce to equation (15.1). In a solid the ligands at $\vec{R}$ are such that, when the sum is performed over the ligands, only certain k and q survive. Much of the above derivation has been given by Judd (1976) in a different context, and many of his elegant techniques could be used to simplify the resulting expressions. For example, using Judd's notation (1976), equation (15.25) becomes

$$U^{(n)}(\vec{r}_1\vec{r}_2,\vec{R}) = \sum_{\substack{ab\\k}} F_n(abk)r_1^a r_2^b [\underline{C}_a(\hat{r}_1)\underline{C}_b(\hat{r}_2)]_k \cdot \underline{C}_k(\hat{R})/R^{a+b+2n+2} \quad,$$

(15.26)

where

$$[\underline{C}_a(\hat{r}_1)\underline{C}_b(\hat{r}_2)]_{kq} = \sum_\alpha \langle a(\alpha)b(q-\alpha)|k(q)\rangle \, C_{a\alpha}(\hat{r}_1)C_{b,q-\alpha}(\hat{r}_2) \quad.$$

The tensor in orbital space, given in equation (15.25),

$$T_{kq}(a,b) = \sum_\alpha \langle a(\alpha)b(q-\alpha)|k(q)\rangle \, C_{a\alpha}(\hat{r}_1)C_{b,q-\alpha}(\hat{r}_2) \quad,$$

(15.27)

should be considered carefully. For a fixed value of k the number of terms in the sum over a and b is restricted by $a + b \leq k$; for equivalent electrons a and b are restricted to even integers; and for $0 \leq (a,b) \leq 6$, the total number of terms is not excessive. But since a and b can reach the maximum value of 6 for the configuration $4f^n$, the value of k in the k sum (similar to the lattice sum) must go up to 12, that is, $k \leq 12$.

If as in previous work (Morrison, 1980) the sum over all the electrons is performed in equation (15.25) along with the sum over the ligands, the results are

$$U^{(n)}(\vec{R}) = 1/2 \sum_{\substack{ij\\R}} U^{(n)}(\vec{r}_i\vec{r}_j,\vec{R}) \quad,$$

(15.28)

where the factor 1/2 accounts for the self-interaction terms that are present when an electron interacts with its own induced multipole, as well as for the interactions that occur twice when $i \neq j$. This interaction contains a large number of corrections to the free-ion parameters, a few of which are discussed in the following.

15.2 Slater Integral Shifts

The Slater integrals for the free-ion interactions are given by the Coulomb interaction as

$$H_1 = \sum_{i>j} \frac{e^2}{|\vec{r}_i - \vec{r}_j|} \quad,$$

(15.29)

148

which for equivalent electrons can be written

$$H_1 = \sum_{k,q} F^{(k)} \sum_{i>j} c_{kq}^*(\hat{r}_j) c_{kq}(\hat{r}_i) \quad , \tag{15.30}$$

where

$$F^{(k)} = e^2 \int_0^\infty \int_0^\infty \frac{r_<^k}{r_>^{k+1}} [R_{n\ell}(r_2)]^2 \, dr_1 dr_2 \quad .$$

Since the interaction represented by equation (15.30) is spherically symmetric in the space of all the electrons, corrections to the $F^{(k)}$ can only arise from terms in an interaction that are spherically symmetric in the space of the electrons. Thus, in equation (15.25) if we let $k = 0$, we have such an interaction, and the following results are achieved:

$$F_n(ab0) = -\alpha_n e^2 \langle a+n(0)a+n(0)|0(0)\rangle \sqrt{(2a+2n+1)(2b+2n+1)}$$

$$\times W(a,a+n,a,a+n;n0) \left[\binom{2a+2n}{2a} \binom{2b+2n}{2b} \right]^{1/2} \tag{15.31}$$

$$= -\alpha_n e^2 (-1)^{a+n} \binom{2a+2n}{2a} / \sqrt{2a+2n+1} \quad .$$

If in equation (15.31) we use the relations

$$\langle a(\alpha)b(-\alpha)|0(0)\rangle = (-1)^{a-\alpha}/\sqrt{2b+1} \; \delta_{ab} \tag{15.32}$$

and

$$W(a,a+n,a,a+n;n0) = (-1)^n/[(2a+1)(2a+2n+1)]^{1/2}$$

(Rose, 1957), then equation (15.25) becomes

$$U^{(n)}(\vec{r}_1,\vec{r}_2,\vec{R}) = -\alpha_n e^2 \sum_a \binom{2a+2n}{2a} \sum_\alpha r_1^a r_2^a \frac{c_{a\alpha}^*(\hat{r}_1) c_{a\alpha}(\hat{r}_2)}{R^{2a+2n+2}} \quad , \tag{15.33}$$

which is the same form as equation (15.30). Thus,

$$\Delta F_n^{(k)} = -\sum_i \alpha_n(i) e^2 \sum_a \binom{2k+2n}{2k} \frac{\langle r^k \rangle^2}{R_i^{2k+2n+2}} \tag{15.34}$$

for the Slater integral shifts, because of the electron multipolar interaction with the ligands of multipolar polarizabilities α_n.

149

15.3 Problems

1. If we have a charge distribution $\rho(\vec{x})$ at the origin, then the electric potential at $\vec{r}$ is

$$d\phi(\vec{r}) = \frac{\rho(\vec{x})d\tau_x}{|\vec{r} - \vec{x}|} \quad ,$$

where $d\tau_x$ is the volume element at $\vec{x}$. By expanding $\frac{1}{|\vec{r} - \vec{x}|}$ for $|\vec{x}| < |\vec{r}|$, show that

$$\phi(\vec{r}) = \sum_{nm} \frac{Q^*_{nm} C_{nm}(\vec{r})}{r^{n+1}}$$

where

$$Q_{nm} = \int \rho(\vec{x}) x^n C_{nm}(\hat{x}) \, d\tau_x$$

is the multipolar moment of the charge distribution. What is the common name for Q_{00} and Q_{1m}?

2. If we consider the charge distribution in problem 1 to be at the point $\vec{r}$, show that the electric potential is

$$\phi(\vec{r}) = \sum_{nm} (-1)^n \frac{Q^*_{nm} C_{nm}(\hat{r})}{r^{n+1}} \quad .$$

3. The electric potential of a uniform electric field $\vec{E}$ is

$$\phi(\vec{x}) = -\vec{x} \cdot \vec{E}$$

or

$$\phi(\vec{x}) = - \sum_m E^*_{1m} x C_{1m}(\hat{x}) \tag{a}$$

when $\vec{E}$ and $\vec{x}$ are written in spherical tensor form.

If we generalize equation (a) above to

$$\phi(\vec{R} + \vec{x}) = - \sum_{nm} E^*_{nm}(\vec{R}) x^n C_{nm}(\hat{x}) \quad , \tag{b}$$

we can define the induced multipole moment caused by a field $E_{nm}(\vec{R})$ as

$$Q_{nm}(\vec{R}) = \alpha_n E_{nm}(\vec{R}) \quad . \tag{c}$$

By taking a point charge q_o at $\vec{R}$, show that the multipolar field at the origin is

$$E_{nm} = -q_o \frac{C_{nm}(\hat{R})}{R^{n+1}}$$

and that

$$Q_{nm}(R) = -q_o \alpha_n \frac{C_{nm}(\hat{R})}{R^{n+1}} \quad . \tag{d}$$

    4.   Using the results of problems 1, 2, and 3, show that the potential at $\vec{r}$ due to the above multipole moments of an ion with polarizability α_n is

$$\phi = \sum_{nm} Q_{nm}^*(R) \frac{C_{nm}(\hat{r})}{r^{n+1}}$$

with $Q_{nm}(R)$ given in the last part of problem 3.

    5.   By considering a dielectric sphere of radius a and dielectric constant ε in the presence of a point charge q_o at $|\vec{r}| > a$, show that the multipole polarizability of the sphere is

$$\alpha_n = [n(\varepsilon - 1)a^{2n+1}]/(\varepsilon n + n + 1) \quad .$$

This result can be obtained by solving the electrostatic problem of a dielectric sphere in the presence of a point charge (Jackson, 1975).

    6.   The energy of interaction of two rigid charge distributions $\rho_A(\vec{x})$ and $\rho_B(\vec{y})$ separated by a distance $\vec{R}$ ($|\vec{R}| \gg |\vec{x}|, |\vec{y}|$) is

$$U_{AB} = \iint \frac{\rho_A(\vec{x}) \, d\tau_x \rho_B(\vec{y}) \, d\tau_y}{|\vec{R} - \vec{x} + \vec{y}|} \quad .$$

By using the two-center expansion (Judd, 1975) on the integrand, show that

$$U_{AB} = \sum_{\substack{a d \\ b \beta}} (-1)^b \binom{2a+2b}{2a}^{1/2} \langle a(\alpha)b(\beta) | a+b(\alpha+\beta) \rangle \, Q_{a\alpha}(A) Q_{b\beta}(B) \frac{C_{a+b,\alpha+\beta}^*(\hat{R})}{R^{a+b+1}}$$

where

$$Q_{a\alpha}(A) = \int \rho_A(\vec{x}) x^n C_{a\alpha}(\hat{x}) \, d\tau_x \quad , \quad \text{etc.}$$

The total energy for a solid consisting of such multipoles would then be the sum over all A and B such that $\vec{R} = \vec{R}_A - \vec{R}_B$.

7. The energy of interaction of two ions, with the first having the electronic configuration $n_A \ell_A^{N_A}$ and the second ion having the electronic configuration $n_B \ell_B^{N_B}$, can be obtained from the result in problem 6 by letting

$$ Q_{a\alpha}(A) = \sum_{i=1}^{N_A} r_i^a C_{a\alpha}(\hat{r}_i) $$

and

$$ Q_{b\beta}(B) = \sum_{j=1}^{N_B} r_j^b C_{b\beta}(\hat{r}_j) \quad . $$

The resulting interaction can be used to calculate energy transfer from ion A to B.

15.4 Annotated Bibliography and References

Carlson, B. C., and G. S. Rushbrooke (1950), On the Expansion of a Coulomb Potential in Spherical Harmonics, Proc. Camb. Phil. Soc. 46, 626. This reference gives a thorough discussion of the two-center expansion used here.

Jackson, J. D. (1975), Classical Electrodynamics, Wiley, New York, NY. Again note the difference of his $q\ell m$ and our $Q\ell m$.

Judd, B. R. (1977), Correlation Crystal Fields for Lanthanide Ions, Phys. Rev. Lett. 39, 242.

Judd, B. R. (1976), Modification of Coulombic Interactions by Polarizable Atoms, Math. Proc. Camb. Phil. Soc. 80, 535.

Judd, B. R. (1975), Angular Momentum Theory for Diatomic Molecules, Academic Press, New York, NY, pp 95-109.

Morrison, C. A. (1980, January 15), Host Dependence of the Rare-Earth Ion Energy Separation $4F^N - 4F^{N-1}n\ell$, J. Chem. Phys. 72, 1001.

Rose, M. E. (1957), Elementary Theory of Angular Momentum, McGraw-Hill, New York, NY, pp 47 and 277.

OVERALL BIBLIOGRAPHY

Ballhausen, C. J. (1962), Introduction to Ligand Field Theory, McGraw-Hill, New York, NY.

Bethe, H. A. (1929), Termaufspaltung in Kristallen (translation: Splitting of Terms in Crystals, by Consultants Bureau, Inc., New York, NY), Ann. Physik (Leipzig) 3, 133.

Bethe, H. (1930), Zur Theorie des Zeemaneffektes an den Salzen der seltenen Erden, Z. Physik 60, 218.

Bishton, S. S., and D. J. Newman (1970), Parametrization of the Correlation Crystal Field, J. Phys. C 3, 1753.

Brink, D. M., and G. R. Satchler (1962), Angular Momentum, Clarendon Press, Oxford, U.K.

Brown, E. A., J. Nemarich, N. Karayianis, and C. A. Morrison (1970, November), Evidence for Yb^{3+} 4f Radial Wavefunction Expansion in Scheelites, Phys. Lett. 33A, 375.

Carlson, B. C., and G. S. Rushbrooke (1950), On the Expansion of Coulomb Potential in Spherical Harmonics, Proc. Camb. Phil. Soc. 46, 626.

Carnall, W. T., H. Crosswhite, and H. M. Crosswhite (1978), Energy Level Structure and Transition Probabilities in the Spectra of the Trivalent Lanthanides in LaF_3, Argonne National Laboratory, ANL-78-XX-95.

Condon, E. U., and H. Odabasi (1980), Atomic Structure, Cambridge University Press, Cambridge, U.K.

Condon, E. U., and G. H. Shortley (1959), The Theory of Atomic Spectra, Cambridge University Press, Cambridge, England.

Cowan, R. D., and D. C. Griffin (1976), Approximate Relativistic Corrections to Atomic Radial Wave Functions, J. Opt. Soc. Am. 66, 1010.

Dieke, G. H. (1968), Spectra and Energy Levels of Rare Earth Ions in Crystals, Interscience Publishers, New York, NY.

Dieke, G. H., and H. M. Crosswhite (1963), The Spectra of the Doubly and Triply Ionized Rare Earths, Appl. Opt. 2, 675.

Edmonds, A. R. (1957), Angular Momentum in Quantum Mechanics, Princeton University Press, Princeton, NJ.

Erdos, P., and J. H. Kang (1972), Electric Shielding of Pr^{3+} and Tm^{3+} Ions in Crystals, Phys. Rev. B6, 3383.

Faucher, M., and P. Caro (1977), A Quickly Converging Method for Computing Electrostatic Crystal Field Parameters, J. Chem. Phys. 66, 1273.

Faucher, M., and D. Garcia (1983a), Crystal Field Effects in Rare-Earth-Doped Oxyhalides: Ab-Initio Calculations Including Effects of Dipolar and Quadrupolar Moments, Solid State Chemistry, Proceedings of the Second European Conference (7-9 June 1982, Veldhoven, Netherlands), Elsevier, Amsterdam, Netherlands.

Faucher, M., and D. Garcia (1983b), Crystal Field Effects on 4f Electrons: Theories and Reality, J. Less-Common Metals 93, 31.

Faucher, M., and D. Garcia (1982), Electrostatic Crystal-Field Contributions in Rare-Earth Compounds with Consistent Multipolar Effects: I.--Contribution to K-Even Parameters, Phys. Rev. B26, 5451.

Fraga, S., K. M. S. Saxena, and J. Karwowski (1976), Physical Science Data 5, Handbook of Atomic Data, Elsevier, New York, NY.

Freeman, A. J., and R. E. Watson (1962), Theoretical Investigation of Some Magnetic and Spectroscopic Properties of Rare-Earth Ions, Phys. Rev. 127, 2058.

Garcia, D. (1983), Simulation ab-initio des parametres du champ des ligandes, thesis, L'Ecole Centrale des Arts et Manufactures.

Garcia, D., and M. Faucher (1984), Crystal-Field Parameters in Rare-Earth Compounds: Extended Charge Contributions, Phys. Rev. A 30, 1703.

Garcia, D., M. Faucher, and O. L. Malta (1983), Electrostatic Crystal-Field Contributions in Rare-Earth Compounds with Consistent Multipolar Effects: II.--Contribution to K-Odd Parameters (Transition Probabilities), Phys. Rev. B 27, 7386.

Griffith, J. S. (1961), The Theory of Transition-Metal Ions, Cambridge University Press, Cambridge, U.K.

Hufner, S. (1978), Optical Spectra of Transparent Rare Earth Compounds, Academic Press, New York, NY.

Hutchings, M. T., and D. K. Ray (1963), Investigations into the Origin of Crystalline Electronic Field Effects on Rare-Earth Ions: I--Contributions from Neighbouring Induced Moments, Proc. Phys. Soc. 81, 663.

International Tables for X-Ray Crystallography (1952), Vol. I, Kynock Press, U.K.

Jackson, J. D. (1975), Classical Electrodynamics, Wiley, New York, NY.

Judd, B. R. (1977), Correlation Crystal Fields for Lanthanide Ions, Phys. Rev. Lett. 39, 242.

Judd, B. R. (1976), Modification of Coulombic Interactions by Polarizable Atoms, Math. Proc. Camb. Phil. Soc. 80, 535.

Judd, B. R. (1975), Angular Momentum Theory for Diatomic Molecules, Academic Press, New York, NY.

Judd, B. R. (1963), Operator Techniques in Atomic Spectroscopy, McGraw-Hill, New York, NY.

Judd, B. R. (1962), Optical Absorption Intensities of Rare-Earth Ions, Phys. Rev. $\underline{127}$, 750.

Karayianis, N. (1971), Theoretical Energy Levels and g Values for the ^{4}I Terms of Nd^{3+} and Er^{3+} in $LiYF_4$, J. Phys. Chem. Solids $\underline{32}$, 2385.

Karayianis, N. (1970, 15 September), Effective Spin-Orbit Hamiltonian, J. Chem. Phys. $\underline{53}$, 2460.

Karayianis, N., and R. T. Farrar (1970, November), Spin-Orbit and Crystal Field Parameters for the Ground Term of Nd^{3+} in $CaWO_4$, J. Chem. Phys. $\underline{53}$, 3436.

Karayianis, N., and C. A. Morrison (1975, January), Rare Earth Ion-Host Crystal Interactions: 2.--Local Distortion and Other Effects in Reconciling Lattice Sums and Phenomenological B_{km}, Harry Diamond Laboratories, HDL-TR-1682.

Karayianis, N., and C. A. Morrison (1973, October), Rare Earth Ion-Host Lattice Interactions. 1.--Point Charge Lattice Sum in Scheelites, Harry Diamond Laboratories, HDL-TR-1648.

Kittel, C. (1956), Introduction to Solid State Physics (2nd ed.), Wiley, New York, NY.

Konig, E., and S. Kremer (1977), Ligand Field Energy Diagrams, Plenum Press, New York, NY.

Koster, G. F., J. O. Dimmock, R. G. Wheeler, and H. Statz (1963), Properties of the Thirty-Two Point Groups, MIT Press, Cambridge, MA.

Leavitt, R. P., and C. A. Morrison (1980, 15 July), Crystal Field Analysis of Triply Ionized Rare Earth Ions in Lanthanum Trifluoride: II--Intensity Calculations, J. Chem. Phys. $\underline{73}$, 749.

Leavitt, R. P., C. A. Morrison, and D. E. Wortman (1975, June), Rare Earth Ion-Host Crystal Interactions: 3.--Three-Parameter Theory of Crystal Fields, Harry Diamond Laboratories, HDL-TR-1673.

Leavitt, R. P., C. A. Morrison, and D. E. Wortman (1974), Description of the Crystal Field for Tb^{3+} in $CaWO_4$, J. Chem. Phys. $\underline{61}$, 1250.

Leighton, R. (1959), Principles of Modern Physics, McGraw-Hill, New York, NY, chapter 5, The One Electron Atom.

Li, Wai-Kee (1971), Reduced Matrix Elements of $V^{(12)}$, $V^{(13)}$, and $V^{(14)}$ for d^n Configurations, Atomic Data $\underline{2}$, 263.

Low, W. (1958a), Paramagnetic and Optical Spectra of Divalent Nickel in Cubic Crystalline Fields, Phys. Rev. $\underline{109}$, 247.

Low, W. (1958b), Paramagnetic and Optical Spectra of Divalent Cobalt in Cubic Crystalline Fields, Phys. Rev. $\underline{109}$, 256.

McClure, D. S. (1959), Electronic Spectra of Molecules and Ions in Crystals, Part II.--Spectra of Ions in Crystals, Solid State Phys. $\underline{9}$, 399.

Merzbacher, E. (1961), Quantum Mechanics, Wiley, New York, NY.

Morrison, C. A. (1980, January 15), Host Dependence of the Rare-Earth Ion Energy Separation $4F^N - 4F^{N-1}n\ell$, J. Chem. Phys. $\underline{72}$, 1001.

Morrison, C. A. (1976), Dipolar Contributions to the Crystal Fields in Ionic Solids, Solid State Commun. $\underline{18}$, 153.

Morrison, C. A., G. F. de Sa, and R. P. Leavitt (1982), Self-Induced Multipole Contribution to the Single-Electron Crystal Field, J. Chem. Phys. $\underline{76}$, 3899.

Morrison, C. A., N. Karayianis, and D. E. Wortman (1977, June), Rare Earth Ion-Host Lattice Interactions: 4.--Predicting Spectra and Intensities of Lanthanides in Crystals, Harry Diamond Laboratories, HDL-TR-1816.

Morrison, C. A., and R. P. Leavitt (1982), Spectroscopic Properties of Triply Ionized Lathanides in Transparent Host Materials, in Volume 5, Handbook of the Physics and Chemistry of Rare Earths, ed. by K. A. Gschneidner, Jr., and L. Eyring, North-Holland Publishers, New York, NY.

Morrison, C. A., and R. P. Leavitt (1979, 15 September), Crystal Field Analysis of Triply Ionized Rare Earth Ions in Lanthanum Trifluoride, J. Chem. Phys. $\underline{71}$, 2366.

Morrison, C. A., R. P. Leavitt, and A. Hansen (1985, October), Host Materials for Transition-Metal Ions with the nd^N Electronic Configuration, Harry Diamond Laboratories, HDL-DS-85-1.

Morrison, C. A., D. R. Mason, and C. Kikuchi (1967), Modified Slater Integrals for an Ion in a Solid, Phys. Lett. $\underline{24A}$, 607.

Morrison, C. A., and R. G. Schmalbach (1985, July), Approximate Values of $\langle r^k \rangle$ for the Divalent, Trivalent, and Quadrivalent Ions with the $3d^N$ Electronic Configuration, Harry Diamond Laboratories, HDL-TL-85-3.

Morrison, C. A., and D. E. Wortman (1971, November), Free Ion Energy Levels of Triply Ionized Thulium Including the Spin-Spin, Orbit-Orbit, and Spin-Other-Orbit Interactions, Harry Diamond Laboratories, HDL-TR-1563.

156

Newman, D. J. (1973), Slater Parameter Shifts in Substituted Lanthanide Ions, J. Phys. Chem. Solids $\underline{34}$, 541.

Nielson, C. W., and G. F. Koster (1963), Spectroscopic Coefficients for the p^n, d^n, and f^n Configurations, MIT Press, Cambridge, MA.

Ofelt, G. S. (1962), Intensities of Crystal Spectra of Rare-Earth Ions, J. Chem. Phys. $\underline{37}$, 511.

Polo, S. R. (1961, June 1), Studies on Crystal Field Theory, Volume I--Text, Volume II--Tables, RCA Laboratories, under contract to Electronics Research Directorate, Air Force Cambridge Research Laboratories, Office of Aerospace Research, contract No. AF 19(604)-5541. [Volume II gives date as June 1, 1961.]

Racah, G. (1942), Theory of Complex Spectra--I, Phys. Rev. $\underline{61}$, 186; II, Phys. Rev. $\underline{62}$ (1942), 438; III, Phys. Rev. $\underline{63}$ (1942), 367; IV, Phys. Rev. $\underline{76}$ (1949), 1352.

Rainville, E. D. (1960), Special Functions, Macmillan, New York, NY.

Rose, M. E. (1957), Elementary Theory of Angular Momentum, Wiley, New York, NY.

Rose, M. E. (1955), Multipole Fields, Wiley, New York, NY.

Rotenberg, M., R. Bevins, N. Metropolis, and J. K. Wooten, Jr. (1969), The 3-j and 6-j Symbols, MIT Press, Cambridge, MA.

Schiff, L. I. (1968), Quantum Mechanics, 3rd ed., McGraw-Hill, New York, NY.

Sengupta, D., and J. O. Artman (1970), Crystal-Field Shielding Parameters for Nd^{3+} and Np^{4+}, Phys. Rev. $\underline{B1}$, 2986.

Slater, J. C. (1960), Quantum Theory of Atomic Structure, Volume II, McGraw-Hill, New York, NY.

Sobelman, I. I. (1979), Atomic Spectra and Radiative Transitions, Springer-Verlag, New York, NY.

Stephens, R. R., and D. E. Wortman (1967, November), Comparison of the Ground Term, Energy Levels, and Crystal Field Parameters of Terbium in Scheelite Crystals, Harry Diamond Laboratories, HDL-TR-1367.

Sternheimer, R. M. (1966), Shielding and Antishielding for Various Ions and Atomic Systems, Phys. Rev. $\underline{146}$, 140.

Sternheimer, R. M. (1951), Nuclear Quadrupole Moments, Phys. Rev. $\underline{84}$, 244.

Sternheimer, R. M., M. Blume, and R. F. Peierls (1968), Shielding of Crystal Fields at Rare-Earth Ions, Phys. Rev. $\underline{173}$, 376.

Tinkham, M. (1964), Group Theory and Quantum Mechanics, McGraw-Hill, New York, NY.

Trees, R. E. (1964), $4f^3$ and $4f^25d$ Configuration of Doubly Ionized Praseodymium (Pr III), J. Opt. Soc. Am. $\underline{54}$, 651.

Uylings, P. H. M., A. J. J. Raassen, and J. F. Wyart (1984), Energies of N Equivalent Electrons Expressed in Terms of Two-Electron Energies and Independent Three-Electron Parameters: A New Complete Set of Orthogonal Operators: II.--Application of $3d^N$ Configurations, J. Phys. $\underline{B17}$, 4103.

Watanabe, H. (1966), Operator Methods in Ligand Field Theory, Prentice-Hall, Englewood Cliffs, NJ.

Wortman, D. E. (1972), Ground Term Energy States for Nd^{3+} in $LiYF_4$, J. Phys. Chem. Solids $\underline{33}$, 311.

Wortman, D. E. (1971a), Optical Spectrum of Triply Ionized Erbium in Calcium Tungstate, J. Chem. Phys. $\underline{54}$, 314.

Wortman, D. E. (1971b), Ground Term of Nd^{3+} in $LiYF_4$, Bull. Am. Phys. Soc. $\underline{16}$, 594.

Wortman, D. E. (1970a), Ground Term Energy Levels and Possible Effect on Laser Action for Er^{3+} in $CaWO_4$, J. Opt. Soc. Am. $\underline{60}$, 1143.

Wortman, D. E. (1970b, June), Optical Spectrum of Triply Ionized Erbium in Calcium Tungstate, Harry Diamond Laboratories, HDL-TR-1510.

Wortman, D. E. (1968a), Analysis of the Ground Term of Tb^{3+} in $CaWO_4$, Phys. Rev. $\underline{175}$, 488.

Wortman, D. E. (1968b), Absorption and Fluorescence Spectra and Crystal Field Parameters of Tb^{3+} in $CaWO_4$, Bull. Am. Phys. Soc. $\underline{13}$, 686.

Wortman, D. E. (1968c, February), Absorption and Fluorescence Spectra and Crystal Field Parameters of Triply Ionized Terbium in Calcium Tungstate, Harry Diamond Laboratories, HDL-TR-1377.

Wortman, D. E., R. P. Leavitt, and C. A. Morrison (1973, December), Analysis of the Ground Configuration of Trivalent Thulium in Single-Crystal Yttrium Vanadate, Harry Diamond Laboratories, HDL-TR-1653.

Wortman, D. E., C. A. Morrison, and N. Karayianis (1977, June), Rare Earth Ion-Host Lattice Interactions: 5.--Lanthanides in $CaWO_4$, Harry Diamond Laboratories, HDL-TR-1794.

Wortman, D. E., and D. Sanders (1971a), Optical Spectrum of Trivalent Dysprosium in Calcium Tungstate, J. Chem. Phys. $\underline{55}$, 3212.

Wortman, D. E., and D. Sanders (1971b, April), Optical Spectrum of Trivalent Dysprosium in Calcium Tungstate, Harry Diamond Laboratories, HDL-TR-1540.

Wortman, D. E., and D. Sanders (1970a), Ground Term Energy Levels of Triply Ionized Holmium in Calcium Tungstate, J. Chem. Phys. $\underline{53}$, 1247.

Wortman, D. E., and D. Sanders (1970b, March), Absorption and Fluorescence Spectra of Ho^{3+} in $CaWO_4$, Harry Diamond Laboratories, HDL-TR-1480.

Wybourne, B. G. (1965), Spectroscopic Properties of Rare Earths, Wiley, New York, NY.

Lecture Notes in Chemistry